The Human Body in Barbarian Laws, c. 500 – c. 800

Przemysław Tyszka

The Human Body in Barbarian Laws, c. 500 – c. 800

Corpus Hominis as a Cultural Category

Translated from the Polish by Guy Russell Torr

Bibliographic Information published by the Deutsche Nationalbibliothek
The Deutsche Nationalbibliothek lists this publication
in the Deutsche Nationalbibliografie; detailed bibliographic
data is available in the internet at http://dnb.d-nb.de.

Library of Congress Cataloging-in-Publication Data
Tyszka, Przemyslaw, 1969-
  The human body in Barbarian laws, c. 500 - c. 800 : corpus hominis
as a cultural category / Przemyslaw Tyszka ; translated from the
Polish by Guy Russell Torr.
    pages cm
  Includes bibliographical references and index.
  ISBN 978-3-631-64230-6 (alk. paper)
  1. Human body–Law and legislation–History 2. Human body–History.
3. Law, Germanic. 4. Law, Medieval. I. Title.
  K564.H8T97 2014
  342.08'5–dc23
                                                     2013043870

This publication was financed by the Institute of History
of the Maria Curie-Skłodowska University in Lublin.

The translation of this work has been funded
by the Foundation for Polish Science.

Cover illustration:
Detail from *Bayeux Tapestry*, c. 1080, Musée de la Tapisserie de Bayeux

Title of the Polish edition:
*Prawa barbarzyńskie o czynach przeciw ciału i cielesności człowieka
(od końca V do początku IX wieku). Corpus hominis jako
kategoria kulturowa*, Wydawnictwo UMCS, Lublin 2010

ISBN 978-3-631-64230-6 (Print)
E-ISBN 978-3-653-03731-9 (E-Book)
DOI 10.3726/978-3-653-03731-9

© Peter Lang GmbH
Internationaler Verlag der Wissenschaften
Frankfurt am Main 2014
All rights reserved.
Peter Lang Edition is an Imprint of Peter Lang GmbH.

Peter Lang – Frankfurt am Main · Bern · Bruxelles · New York ·
Oxford · Warszawa · Wien

All parts of this publication are protected by copyright. Any
utilisation outside the strict limits of the copyright law, without
the permission of the publisher, is forbidden and liable to prosecution. This
applies in particular to reproductions, translations, microfilming, and storage and processing in electronic retrieval systems

www.peterlang.com

Contents

Acknowledgments 7

Introduction 9
The Problem Area of the Work 9
The State of Research into the Question of the Human Body
in Germanic Laws 14
Research Methods into Barbarian Laws 23
A Review of the Questions Examined in the Work 25

Chapter One
**The Law Books of Germanic Peoples as a Source for Research
into the Human Body** 29
The Origin, Character and Functions of the Written Laws of the Germanic
Peoples 29
The Problem of the Relationship between the Recording of Law
and the Oral Legal Tradition 40

Chapter Two
**The Concept of the Human Body in the Germanic Legal Codes
and in Early Medieval Narratives** 59
The Concept of the Human Body within the Barbarian *Leges* 60
Excursus on the Use of the Word *Membrum* 69
The Specifics of Understanding the Concept of 'Body' in Visigoth
and Lombard Law 72
The Means by Which the Collection of Regulations on Crimes against
Parts of the Body Were Composed. 81
The Factors Forming the Way of Presenting the Structure of the Human
Body in the *Leges* 94
    Barbarian *Leges* – A Type of Utterance on Man's Body
    as the Object of Crime 96
    Barbarian *Leges* and Forms of Thought Transfer in an Oral Culture 98
    The Human Body in Christian Eschatology and the Pagan Beliefs
    of the Germanic Peoples 106

Chapter Three
The Human Body as an Object of Crime. Body Parts, Their Damage and Violation and the Compensatory Tariff System 115

Body Parts as Objects of Crime – an Analysis of the Content of Selected *Leges* 116

Body Parts – an Overall Look at the *Leges* Collections 123

Types of Crime against the Body 127

Descriptions of Crimes against Bodily Integrity as an Account on the Human Body 142

The Effects of Crimes against the Body 146

Compensatory Systems and the Problem of Body Part Evaluation 154

    The Origin of the Compensatory System for Violation of the Human Body 157

    Tariff Payments for Bodily Injuries 160

    The Arbitrary and Comprehensive Nature of the Compensation Systems ... 164

Chapter Four
The Perception of the Human Body and Social Differentiation amongst Germanic Peoples 171

The Reasons for Social Differentiation in the Values of Compensation for Crimes against the Body 174

Images of Germanic Social Structure and the View of the Human Body as Perceived by the *Leges* Codifiers 181

    'Images' of the Social Structure of Germanic Peoples – a Comparative Analysis of Selected Examples. 182

    Textual Presentations of the Human Body and Model Depictions of Social Structure 199

    The Human Body and the Hierarchical Structure of Social Groups 209

Man's Body and Affiliation to the Society 211

Conclusion 221

Abbreviations used in the footnotes and bibliography 233

Bibliography 235

Acknowledgments

This work arose within the framework of a Polish Ministry of Science and Higher Education research project *Kulturowy obraz ciała ludzkiego we wczesnośredniowiecznych spisach praw ludów germańskich* [The Cultural Image of the Human Body in the Early Medieval Legal Codes of Germanic Peoples]. Its creation was brought about by a study grant awarded by the Netherlands Institute for Advanced Study and the Lanckoronski Foundation, thanks to whom it was possible to conduct preliminary research in foreign libraries. Of immense assistance was also the financial support I was granted by the authorities of the Humanities Faculty and the Institute of History at the Marie Curie-Skłodowska University in Lublin. I would like to express my deep gratitude to all the above mentioned institutions. I would also like to thank Professor Jacek Banaszkiewicz for his inspiring comments and pointers on various aspects of the material herein researched. I would like to thank Professor Dariusz Słapek, Director of the Institute of History of the Maria Curie Słodowska University, for his kind help in publishing this book.

Introduction

The Problem Area of the Work

The subject of the undertaken research is the body and corporality in the early medieval legal codes of Germanic peoples (*leges barbarorum*). Such an area of interest requires an explanation and an exact definition. The matter concerns, first and foremost, the very category of the human body. This is connected with extremely important questions for our area of interest: in what meaning are we able to treat the regulations that appear in *leges barbarorum* as an account of the human body? What possibilities within this are offered to the historian researching the human body and this type of source? What limitations and restrictions result from the specifics of these accounts for the problem area in question?

The first factor characterising the concepts dealt with in the research is the fact that the subject of the considerations is not a 'real', physical human body but its textual representation as appearing in *leges barbarorum*. This ascertainment although appearing to be rather obvious in nature has, however, far reaching consequences. For it means that in our case the body is not the object undergoing direct investigation (e.g., observation, reconstruction on the basis of bone remains etc.) as is possible in archaeological research. In the regulations comprising barbarian laws we are dealing with a specific way of presenting selected fragments of man's bodily exterior.

The most important source of knowledge about the human body in the *leges* of Germanic peoples are the regulations on varied crimes against the integrality and inviolability of the human body. The object of the said acts of violence were specified parts or elements of the body. In particular barbarian codes we can find greater or lesser collections of these types of bodily objects. These regulations in which they appear concerned almost exclusively the body of a live man. In *leges* we do also have references to regulations on corpses and the dead.

The fundamental research problem is linked with the said regulations. The above presentation of the content of the *leges* regulations is a certain interpretation of them resulting from reading the said regulations, in accordance with the contemporary understanding of the concept of 'the body'. In the majority of the codifications researched the said fragments of the human body were not linked to the concept of the body. Today's reader understands an account talking, for example, about the cutting off of a hand as a mention of one of the parts of the body. He will assume therefore that legal regulations concerning this type of crime will contain information about the body. Such a comprehended concept of the body is abstract and autonomic in relation to the concept of man. The contemporary researcher adopted more or less consciously that the body is an objective and transhistorical category (and even ahistorical) which may be applied in research into source accounts coming from any epoch whatsoever without taking into consideration its cultural changeability over the course of history.

A man's body is, however, also a historically variable cultural category. One may assume as highly likely that in early medieval Germanic culture, whose manifestation are the *leges*, the understanding of the concept of the body was somewhat different than today's. An important aim of the research into barbarian codifications is therefore to determine in what way their creators understood this concept.

Interesting considerations, from the point of view of the problems herein addressed, on the way the concept of the body was understood in ancient culture, on the basis of research into the language of the works of Homer, have been made by Bruno Snell. Homer used the word σωμά in the meaning of corpse, and not in the sense of the living body that Ancient Greek was to later formulate. Snell also analyses the use of other words appearing in the language of Homer. These are the words γυια and μελεα, representing the members that he used in the meaning of 'body'. The cited scholar draws attention here to the use of the plural of these words and not the singular as is the accepted practice in the contemporary language. This author analyses also the use of the word 'chros', which in his opinion meant 'skin'. He notes, however, that certain researchers have sided with the word meaning 'body'. This is not, however, skin in the anatomical meaning (*derma*), but the surface, the external boundary of the human form. The word 'chros' was used, however, by Homer in the meaning of 'body'. Snell establishes that it would be difficult for a contemporary man to comprehend a mentality that did not recognise the concept of body as such.[1] Snell's findings show, on the basis of a concrete example, that the way of comprehend-

---

1   Cf. B. Snell, *The Discovery of the Mind in Greek Philosophy and Literature*, transl. T. G. Rosenmeyer, New York 1982; Chapter I: *Homer's view of the man*, pp. 1–7; the first German edition: *Die Entdeckung des Geistes: Studien zur des europäischen Denkens bei den Griechen*, Hamburg 1946. Snell reconstructs the evolution of the perception of the body in the epoch researched by him in the following way: in the first stage a speaker standing opposite another person needed in a straightforward way to call him by name: this is Achilles, or the claim: this is a man. Next are described the most distinguishing elements of his appearance, namely his limbs as existing one next to another while their mutual functionality started to be perceived in its full sense only somewhat later. Such an objectively comprehended body was not to exist until the man was not determined by a word, and therefore only when he became the subject of reflection.

ing the concept of the body is connected with the stage of cultural development of the society using it.

Key for our research into the regulations of the *leges* is the question therefore of the relation between the category of person (man) appearing within them and the collections of parts and also fragments of a body. In the barbarian legal codes the said collections were often created somehow on the model of the human figure viewed from head to foot. There arises therefore the question as to whether the creators of the *leges* of the particular Germanic peoples combined the said parts with the concept of the body or with the concept of man. That which a modern researcher identifies as a part of the body, the creators of the *leges* might have understood as objects belonging to a man or also combining to form his figure. It follows to note, however, that man within the researched regulations is perceived in an extremely specific and narrow manner – namely as a bodily being. For he appears in the role of the victim of physical violence. In effect, man appears within the light of the *leges* regulations as the collection of body parts and the motor and perception functions connected with them. It seems therein that the element which connected the said collection of body parts with the concept of a man was his figure.

Within the research herein conducted it is not possible to avoid the use of the concept of 'a man's body' in its modern understanding. An attempt to identify the researcher with the way of thinking inherent for the creators of the sources (in this case barbarian *leges*) would be not only practically unrealisable but would also form a limitation in interpretative possibilities. Therefore it appears that the best solution is the application in this research of the modern concept of the body, with the awareness of the limitations of this approach, as equally a search for the historical meaning of this category.

The second matter requiring specification is the contexts within which the body appears (in the modern understanding) in the *leges* regulations. The human body, or rather its concrete fragments, is – in the light of our source material – first and foremost the object of varied crimes (woundings, damage, violations). It is presented in an extremely literal way – in the form of skin, bones, tendons, innards, blood, hair etc. The body has, therefore, first and foremost a physical or physiological character. The said body 'ingredients' are in the *leges* examined and presented as the subject of an autopsy and therefore in comprehending the matter within modern categories from the viewpoint of forensic medicine.

The body so understood is simultaneously the subject of legal regulations. It becomes the subject of law and as such it is perceived. For the creators of the said regulations were interested in the type and scope of the bodily damages brought about by crime. They were concerned with defining the degree to which a concrete form of damage negatively affected a man's health situation, motor and perceptual ability. Crimes with the use of violence did not always result in damage to the bodily substance, in certain cases the matter concerned violation of bodily inviolability. Violation of goods was consequently not merely the body in the physical sense but the victim's dignity and honour invested in the body. Therefore a man's body was perceived as constituting moral value.[2] This concerned also crimes violating the substance of the body, particularly those whose outcome was a ridiculing change in appearance of the victim. In the light of the regulations of Germanic laws the body constituted a collective of an individual's rights which were subject to protection.

In research into the *leges* it appears justified to use beside the category of 'body' the concept of 'corporality'. In the case of certain types of crime the permanent effects that are their consequence concerned not only the substance of the body but certain functions such as: sight, hearing, speech, the ability to procreate. These were inseparably linked to the existence and functioning of certain organs (the eyes, tongue, genitals), but they themselves were something more. The concept of corporality covers therefore categories connected with the body but extending beyond its strictly material or physical nature (e.g., in as far as an eye is a typical body component, then sight belongs to the sphere of corporality). This category also included nakedness and body appearance.

The sets of body parts appearing in the *leges* were presented as objects defined in social terms. Each fragment of a crime victim's body to appear in the regulations of the said legal codes was viewed from the standpoint of its legal and social status. Victims were characterised in relation to their membership of: groups ('estates') of a specified status (freeman, dependants, slaves), sex and age. Inclusion in one of the social categories decided on the way the sums were estimated that were to be paid by the perpetrator of the crime. In this way a human's body was connected with legal-social differentiation and at the same time was included within the social hierarchical order.

In the light of the *leges* regulations, a human's body covered several aspects: it had a physical or physiological character, it constituted the subject of legal regulations, whose aim was the protection of the individual's property, it was defined in social categories in relation to the status of the victim of the crime. This specific way of viewing the human body meant that in the *leges* regulations there did not appear aspects of it such as: symbolic meanings, beauty, hygienic beautifying procedures, pleasure and suffering etc. The scope and means in

---

2   On the subject of a man's honour and the connection of this category with the body in Germanic and Celtic societies of the early Middle Ages recently there has appeared: J. H. M. Smith, *Europe after Rome. A New Cultural History 500-1000*, Oxford 2005, pp. 100 ff.; cf. also the literature in the bibliography for Chapter 3.

which it is present in the *leges* was strictly connected with the way of understanding reality in the normative texts.

The sets of parts or body elements that we come across in the individual *leges* are in principle presented to some extent 'from outside', for they always create the bodily shell of another person. The body is at the same time devoid of all individual features. It does not belong to any actual person. It is defined socially, ascribed to the legal status of an individual, but it has the character of a 'species', it is a general human trait.

* * *

The source material for our considerations is first and foremost the preserved legal codes of the relevant Germanic peoples and strictly speaking those of the regulations which concern crimes against the body.[3] This is an extensive collection of sources. Our research will concern first and foremost the oldest editions of the *leges* of the particular barbarian peoples. This will involve consequently only a slight recourse or even omitting of some of the later codifications.[4] Such a selection results primarily from the intention to research the process of transition from a traditional culture based on oral transfer and one remaining under the influence of pagan beliefs to a culture of the written word characterised by clear Roman-Christian influences.[5] For we shall treat the barbarian *leges* not on-

---

3   These are: *Leges Visigothorum* (Leovigil's *Antiqua* and Reccesvinth's *Liber Iudiciorum*), the Burgundian *Liber Constitutionum*, *Pactus legis Salicae* and *Lex Salica*, *Lex Riburia*, the Anglo-Saxon (Kentish) the *Laws of Æthelberht*; the Lombard *Edictum Rothari*; *Pactus legis Alamannorum* and *Lex Alamannorum*, *Lex Baiwariorum*; *Lex Saxonum* and *Capitulatio de partibus Saxoniae*, *Lex Thuringorum* and *Lex Frisionum* (including *Additio sapientum*); detailed information about the editions of the various codes is given in the Bibliography and footnotes. The presently available source editions (published in the series *MGH Leges*) were, starting from the beginning of the 20th century, to be the subject of critical evaluation by reviewers; recently W. Hartmann (*Brauchen wir neue Editionen der Leges?* [in:] *Mittelalterlische Texte. Überlieferung – Befunde - Deutungen*, ed. R. Scheiffer (MGH Schriften, vol. 42), pp. 233-245) has correlated these opinions, added his own critical comments and pointed to the need for analysis of the new editions of Alamann, Bavarian, Burgundian and Lombard laws as a result of the errors and absences occurring in hitherto editions, he has also critically evaluated the edition of the Salian Franks by K. A. Eckhardt. As a result of the lack of other possibilities (particularly the ability to make use of manuscripts) I have based myself within the present work on the existing, imperfect editions of the Germanic codes.

4   In the case of the Salian Franks we will not involve ourselves with the capitularies of the Caroline period, while those from the Merovingian era only in passing; in relation to Lombard laws the *Liutprand Laws* we will use only selectively; in the case of Anglo-Saxon legislation we will not make use of research into the later codifications, in particular the Laws of Alfred.

5   For each of the early medieval Germanic peoples this phenomenon occurred at a different historical time, for example in the history of the Visigoths at the turn of the 6th cen-

ly as material for research into the history of law but first and foremost as a certain type of textual utterance containing a cultural account of, among other things, the perception of the human body.

Beside the texts of the *leges* themselves we have used in certain cases, as comparative material, other sources. These are codifications of Roman law (*Codex Theodosianus, Digesta Justiniani, Breviarium Alaricianum, Lex Romana Burgundionum*); early medieval accounts of a historiographic and hagiographic nature (*The Histories* of Gregory of Tours), biographical works (*The Life of Charlemagne* by Einhard, *Epistulae* by Sidonius Apollinaris), mythological accounts (*The Song of Rig*).

The chronological-territorial framework for the work covers the time in which the legal codes came about, while the land area that inhabited by Germanic peoples, for whom the texts were written. This is therefore a period which may be only demarcated by poorly defined starting and finishing points – from the turn of the 6th to the beginning of the 9th century. Equally vaguely may the spatial scope of the work be defined – for here the matter concerned almost the whole of western Europe, and more precisely those areas under the control of the Visigoths, the Franks (Salian and Ripuarian), the Burgundians, Lombards, Bavarians, Alamanns, Thuringii, Saxons, Frisians, and Anglo-Saxons.

The State of Research into the Question of the Human Body in Germanic Laws

Shortly after the publication of the Polish version of my work there was published Lisi Oliver's book *The Body Legal in Barbarian Law*, which covered almost the self same problem area. This monograph is a holistic account of the topic therein undertaken. The author claims that the main topic for her study is 'the personal injury tariffs included in the legal codes established for the various continental kingdoms and dukedoms (...) and incorporating the Anglo-Saxon island regions of Britain'. The thematic scope of her research covers four main problem groups: 'the causes and results of injuries inflicted in private altercation; the evidence for methods and successes (or lack thereof) of healing techniques; the process of individual redress or public litigation; and indication of how the early medieval laws either drew upon native tradition, borrowed regula-

---

tury or even somewhat earlier, in the case of the Frisians at the beginning of the 9th century; hence in relation to the whole group we may observe the process through the course of over three centuries.

tion from other regions, or innovated when need arose in particular cases.'.[6] The author makes use in her research of not only the regulations of barbarian laws but also archaeological, narrative and literary sources. The aim of her study is 'to depict a picture of how early medieval society understood the anatomical consequences of wounds to the human body, the varieties of practices available to heal such injuries, and the legal process of obtaining compensation for temporary or permanent incapacitation'. This book, with regard to its subject matter and comprehensive character, is a contribution to research into the perception of the human body in the early Middle Ages which can simply not be overlooked.[7]

Earlier research into barbarian laws concentrated on questions connected with the history of criminal law.[8] Their main subject was the various types of crime against the body as well as the forms of punishment meted out in the case of them being committed. The human body as a separate problem was one that did not interest the authors of these works. In addition these studies were cross-sectional in nature – they covered equally historical periods of the later rather than earlier Middle Ages. In those works that have come into being during the course of the last thirty years researchers' attention has been concentrated to a lesser degree on crimes against bodily integrity as a historical-legal issue, with greater attention being paid to the various forms of damage inflicted on the human body, and through this on the very body as the object of interest of legal texts.[9] Of special interest for us are the studies conducted by Anette Niederhell-

---

6 L. Oliver, *The Body Legal in Barbarian Law*, Toronto-Buffalo-London 2011, p. 3.
7 The Polish version of the present book was published in September 2010 meaning that I can relate to Lisi Oliver's theses and conceptions, which appeared in May 2011, only within the framework of the English version of the work from the viewpoint of my own earlier independently formulated and realised conceptions into research on the human body as a subject of interest for barbarian laws.
8 W. E. Wilda, *Das Strafrecht der Germanen*, Halle 1842; L. Günther, *Ueber die Hauptstadien der geschichtlichen Entwicklung des Verbrechens der Körperverletzung und seiner Bestrafung*, iur. Diss., Erlangen 1884; R. His, *Die Körperverletzungen im Strafrecht des deutschen Mittelalters*, ZRG GA 41, 1920, pp. 75–126. K. von. Amira, *Die germanischen Todesstrafen*, München 1922; S. von Schwanenflügel, *Die Körperverletzung in der ersten geschriebenen Rechten der Germanen (etwa 500–1300 n. Chr.)*, iur. Diss. (masch.) Göttingen 1950.
9 A. Niederhellmann, *Arzt und Heilkunde in der frühmittelalterlichen Leges*, Berlin 1983; cf. also, eadem, *Heilkundiches in den Leges. Die Schadelverletzungen und ihre Bezeichnungen*, [in:] *Wörter und Sachen im Lichte der Bezeichnungsforschung*, red. R. Schmidt-Wiegand, Berlin–New York 1981, pp. 74–90; M. Elsakkers, *Inflicting serious bodily harm: the Visigothic 'Antiquae' on violence and abortion*, Journal of Legal History, vol. 71/1 (2002/2003), pp. 55–63; eadem, *Abortion,poisoning, magic, and contraconception in Eckhardt's 'Pactus Legis Salicae'*, [in:] *Quod vulgo dicitur. Studien zum*

mann, who, in utilising earlier works (of L. Günther, R. His, K. von Amira, S. von Schwanenflügel), presented a detailed analysis and classification of the particular types of bodily damage described in the individual barbarian *leges* from the viewpoint of their medical character.[10] Other research into the said legal codes have covered also the question of the level of compensation for various types of damage and bodily violation.[11] The subject of interest for historians researching the laws of barbarian peoples is equally the structure of those parts of the codes that contain regulations on crimes against the body.[12]

A different research direction was adopted by Michel Rouche in a text devoted to early medieval Gaul, included in the first volume of *History of a Private Life*.[13] The author treats the human body as an element of the problem of

---

Altniederländischen, red. W. Pijnenburg, A. Quak, T. Schoonheim, Amsterdam– New York 2001, pp. 233–267; eadem, *Genre Hopping: Aristotelian Criteria for Abortion in Germania*, [in:] *Germanic Texts and Latin Models. Medieval Reconstructions*, eds. K. E. Olsen, A. Harbus, T. Hofstra, Luewen–Paris–Sterling 2001, pp. 73–92; L. Oliver, *The Beginnings of English Law*, Toronto 2002, pp. 99–105.

10 The author compared and analysed the regulations of the *leges* with regard to the following phenomena: pregnancy, abortion, contraception, castration; the Germanic (i.e. Burgundian, Frankish, Lombard, Alamann, Bavarian, Saxon) designations for: parts of the body, parts of the head and torso, the limbs, bodily secretions, designations for damage to the body in general, the skin, swellings, bloody damage, bone damage, damage to the skull, torso and internal organs, terms for bodily deformation and inertia, manifestations of the consequences of bodily damage.

11 N. McLeod, *Parallel and paradox. Compensation in the legal systems of Celtic Ireland and Anglo-Saxon England*, Studia Celtica, vol. 16/17 (1981–1982), pp. 25–72; idem, *Compensation for Fingers and Teeth in Early Irish Law*, Peritia, vol. 16 (2002), pp. 344–359; S. Rubin, *The 'Bot', or compositon in Anglo-Saxon Law: A Reassessment*, The Journal of Legal History, vol. 17, no. 2 (1996), pp. 144–154; H. Nijdam, *Measuring wounds in the 'Lex Frisionum' and the Old Frisian Register of Fines*, Philologia Frisica, 1999, pp. 180–203.

12 P. Wormald, *'Inter cetera bona ... gentis suae': law-making and peace-keeping in the earliest English kingdoms*, [in:] *La giustizia nell'alto medioevo (secoli V–VIII)*, vol. II, ed. C. Leonardi, Spoleto 1995, pp. 963–996; idem, *The Making of English Law: King Alfred to the Twelfth Century*, vol. 1, *Legislation and Its Limits*, Oxford 1999; idem, *The leges barbarorum: law and ethnicity in the post-roman West*, in: *Regna and Gentes: The relationship between Late Antiquity and Early Medieval Peoples and Kingdoms in the Transformation of the Roman World*, eds. H.-W. Goetz, J. Jarnut, and w. Pohl, Leiden-Boston 2003, pp. 21-53; L. Oliver, *op. cit.*, pp. 35–38; G. Ausenda, *Jural relations among the Saxons before and after Christianization*, [in:] *The Continental Saxons. From the Migration Period to the Tenth Century: An Ethnographic Perspective*, ed. D. H. Green, F. Siegmund, San Marino 2003, pp. 113–131.

13 M. Rouche, *Haute Moyen Age occidental*, [in:] *Histoire de la vie privée*, vol. I: *De l'Émpire Romain á l'an mil*, ed. P. Veyene, Paris 1985, pp. 467-499; English edition: *A

privacy and family life within Germanic and Roman societies. The *leges* of the Salian, Ripuarian Franks, Burgundians, Visigoths and other peoples served him as one of the sources to research the problem of violence, sexuality, procreation and attitudes towards the dead. A different and one should say innovative approach characterises the study by Katherine O'Brien O'Keeffe on the body in Anglo-Saxon laws of the 9th to the 11th century.[14] The material for this research are the regulations of the later Anglo-Saxon codes on bodily punishments (mutilations) and ordeals. The body is understood in this depiction as the object of the action of secular (convicts) or ecclesiastical (penitents) authority as well as of Divine intervention (miraculous release from suffering).

Even though these works are valuable for our investigations we are still entering into an area that is relatively poorly explored by contemporary medieval studies. The studies where the question of the body has been undertaken concern, with only a few exceptions, either the early Christian periods, the furthest being the 5th century, or the later Middle Ages, starting from the 13th century[15]; while in those devoted to the early Middle Ages, this problem area is usually simply an aside. Besides, it follows to emphasise that both the subject under consideration as well as the source material used make for a complex undertaking.

* * *

The codes of barbarian law as historical documents are specific in character – for they are normative sources. This type of historical account in itself constitutes a complicated, multi-aspectual problem area requiring the application of appropriate interpretative methods. This results from, among other things, the fact that the relations between the text and the historical reality it relates to is formulated within legal texts in a most specific way, differently than in the case of narrative accounts (historiographical or hagiographic) and diplomatic ones. For the legal regulations expressed the convictions of their creators as to the proper and improper behaviour of those subjected to them. Besides, the texts of the *leges* serve in our case to investigate the human body, which equally constitutes a separate and complex problem in itself. The combination of both of these questions examined within the scope of a single study increases the range of difficulty faced by the researcher.

The codifications of Germanic laws as a research subject has the mid nineteenth century as its beginnings. These are first and foremost studies in the histo-

---

    History of Private Life, vol. 1: *From Pagan Rome to Byzantium*, transl. by A. Goldhammer, Cambrigde, Mass. 1992, pp. 453-518.
14 K. O'Brien O'Keeffe, *Body and law in Anglo-Saxon England*, Anglo-Saxon England, 27 (1998), pp. 209–232.
15 Cf. footnotes 34–37.

ry of law, ones written chiefly in German-language works, although there are examples from English, Spanish, Italian and Dutch academic circles.[16] We shall point out here only some of the works from the extensive academic research that has occurred over the course of this time. A huge development in research of this type occurred in Germany during the activity of the *Rechtsschule*, which today is referred to as a classic school of law history, towards the end of the 19th century and during the first decades of the twentieth.[17] Studies into the laws of Germanic peoples were continued over the course of the subsequent decades. It follows here to recall in particular the works of Karl August Eckhardt, the author of the edition of *Lex Salica*[18] and other codes of barbarian laws, as equally Ruth Schmidt-Wiegand.[19] In the 1970s there arose in German-language subject literature a new research current into the laws of Germanic peoples. Here need to be

---

16   A review of the German research: C. Schott, *Der Stand der Leges-Forschung*, Frühmittelalterliche Studien, vol. 13, 1979, pp. 29–55; cf. also, W. Sellert, *Aufzeichnung des Recht und Gesetz*, [in:] *Das Gesetz im Spatantike und frühem Mittelalter*, ed. W. Sellert, Göttingen 1992, pp. 67–102, who presents a critical account of the German literature on the subject from the 19th and 20th century as well as P. Wormald, *'Lex Scripta' and 'Verbum Regis': Legislation and Germanic Kingship, from Euric to Knut*, [in:] *Early Medieval Kingship*, ed. P. H. Sawyer, I. N Wood, Leeds 1977, pp. 105–138; R. Collins, *Visigothic Spain 409-711*, Oxford 2004, cites Spanish literature on the matter, including recent works; an overview of older studies, chiefly German, is contained in the entries for individual legal codes for the specific Germanic peoples, [in:] *Handwörterbuch zur deutschen Rechtsgeschichte*, [hereafter: *HRG*] hg. von A. Ekler, E. Kauffman, R. Schmidt-Wiegand, vol. II, Berlin 1978.

17   Cf. particularly the works of H. Brunner: *Deutsche Rechtsgeschichte*, vol. I, Leipzig 1906 [2nd edition], vol. II Leipzig 1887 [1st edition] as well as idem, *Abhandlungen zur Rechtsgeschichte*, Weimar 1931; K. Zeumera, *Geschichte der westgotischen Gesetzgebung I*, Neues Archiv, vol. 23 (1898) pp. 419–516, 24 (1899) pp. 39–122 and 571–630, 26 (1901) pp. 91–149; K. von Amira, *Grundriss des germanischen Rechts*, Strassburg 1901, idem, *Die germanischen Todesstrafen*, München 1922; F. Beyerle, *Über Normtypen und Erweiterungen der Lex Salica*, Zeitschrift zur Rechtsgeschichte, Germansitische Abteilung, vol. 40 (1924), pp. 216–261; idem, *Die beiden süddeutschen Stammesrechte*, ZRG GA, vol. 73 (1956), pp. 84–140.

18   Cf. K. A. Eckhardt, *Zur Enstehungzeit der Lex Salica*, [in:] *Festschrift zur Feier der 200jährigen Bestehens d. Akademie der Wissenschaften in Göttingen*, Göttingen 1951, pp. 1–31; introductions to *Pactus legis Salicae* and *Lex Salica*.

19   Particularly: R. Schmidt-Wiegand, *Zur Geschichte den malbergischen Glossen*, ZRG GA, vol. 74 (1957), pp. 220–231; eadem, *Das frankische Wortgut der Lex Salica als Denkmal des Westfrankischen*, Rheinsche Vierteljahrblätter, vol. 33 (1969), pp. 396–422; eadem, *Stammesrecht und Volkssprache*, ed. D. Hüpper, Weinheim 1991, contains a collection of the author's works from the end of the 1950s to the end of the 1980s.

enumerated the works of Hermann Nehlsen, Harlad Siems, Clausdieter Schott and Hanna Vollrath.[20] In their works these researchers have on the whole involved themselves in the codes of laws of the Germanic kingdoms in relation to their specific features as textual accounts, their significance for court practice as well as their relationship to the oral transfer of the law, while to a lesser degree and in a different way than earlier historians they have treated it as material for the reconstruction of legal, political-administrative and social reality.

The historical-legal research conducted by British researchers at the end of the nineteenth century involved itself initially in Anglo-Saxon laws.[21] Studies into the laws of Germanic peoples inhabiting continental Europe started to develop a wider academic interest in the 1950s thanks to the work of John M. Wallace-Hadrill.[22] Its significant development was witnessed, however, only with the start of the 1970s. Here mention should be given to the works of Patrick Wormald, Roger Collins, Rosamond McKitterck and Ian Wood.[23] In addition,

---

20 H. Nehlsen, *Skalvenrecht zwischen Antike und Mittelalter. Germanisches und römisches Recht in der germanischen Rechtsaufzeichnungen I. Ostgoten, Westgoten, Franken, Langobarden*, Göttingen 1972; idem, *Zur Aktualität und Efektivitatät der germanischer Rechtsaufzeichnungen*, [in:] *Recht und Schrift im Mittelalter*, ed. P. Classen, Sigmaringen 1977, pp. 449–502; H. Siems, *Studien zur Lex Frisionum*, Ebelsbach am Main 1980; C. Schott, *Pactus, Lex und Recht*, [in:] *Die Alemannen in der Frühzeit*, ed. W. Hübner, Bühl-Baden 1974, pp. 135–168; idem, *Zur Geltung der Lex Alamannorum*, [in:] *Die historische Landschaft zwischen Lech und Vogesen; Forschungen und Fragen zur gesamtalemannischen Geschichte*, ed. P. Fried,W.-D. Sick, Freiburg 1988, pp. 75–105; H. Vollrath, *Gesetzgebung und Schriflichkeit. Das Beispiel der angelsächsischen Gesetze*, Historische Jahrbuch, vol. 99 (1979) pp. 28–54.

21 Cf. F. Pollock, F. W. Maitland, *The History of English Law*, vol. 1, Cambridge 1895; H. G. Richardson, G. O. Sayles, *Law and Legislation*, Edinburgh 1965.

22 J. M. Wallace-Hadrill, *The Long-Haired Kings*, London 1962, particularly Chapter V. *Archbishop Hincmar and the authorship of Lex Salica*, pp. 95–120 (initially published in: Revue d'histoire du droit, vol. XX, 1952) as well as Chapter VI. *The bloodfeud of the Franks*, pp. 121–147 (initial print: Bulletin of the John Rylands Library, vol. 41/2, 1959); idem, *Early Germanic Kingship in England and on the Continent*, Oxford 1971.

23 P. Wormald, *'Lex scripta'*..., passim; idem, *'Inter cetera'*..., passim; idem, *The Making*..., passim; R. Collins, *Early Medieval Spain. Unity in Diversity, 400–1000*, London 1995, pp. 24–31; idem, *Law and ethnic identity in the Western Kingdoms in the fifth and sixth centuries*, [in:] *Medieval Europeans: Studies in Ethnic Identity and National Perspectives in Medieval Europe*, ed. A. P. Smyth, Basingstoke 1998, s. 1–23; idem, *Visigothic*..., pp. 223-239; R. McKitterick, *Some Carolingian law-books and their function*, [in:] *Authority and Power. Studies on Medieval Law and Government Presented to Walter Ullman on His Seventieth Birthday*, eds. B. Tierney, P. Lienehan, Cambridge 1980, pp. 13–27; idem, *Carolingians and the Written Word*, Cambridge 1989, pp. 23–75; I. Wood, *Administration, law and culture in Merovingian Gaul*, [in:] *The Uses of Literacy*

studies into barbarian legal codes have been conducted in the United States. These are first and foremost works by Katherine Fischer Drew and Lisi Oliver.[24] Research into the law of the Germanic peoples was conducted in Italy and Spain, initially in the 19th century, and is at present being continued. The Italian works deal with the Ostrogoths and first and foremost the Lombards as well as the relations between them and general Roman law. Much attention has been devoted to *Edictum Rothari*.[25] Of the research conducted in the 1950s and 1960s it follows to mention the works of G. P. Bognetti and B. Paradisi, who initiated analytical studies into the said edict.[26] In the course of the last three decades there have occurred studies in which the codes of Lombard law have been examined as sources for the history of culture; here the matter concerns the works of Paolo Delogu, Stefano Gasparri and Claudio Azarra.[27]

In Spanish literature on the subject, both that of an older date and the contemporary, the overwhelming emphasis in works on the codifications of Germanic laws is research into *leges Visigothorum*. These works are historical-legal

---

in *Early Medieval Europe*, ed. R. McKitterick, Cambridge 1990, pp. 63–81; idem, *Merovingian Kingdoms 450–751*, London–New York 1994, pp. 102–119; L. Oliver, *Beginning...*, passim; eadem, *Body Legal...*, pp. 8-25.

24 K. Fischer Drew, *Introduction*, [in:] *The Burgundian Code. Book of Constitutions or Law of Gundobad. Additional Enactments*, transl. K. Fischer Drew, Philadelphia 1972, pp. 1–10; eadem, *Introduction*, [in:] *The Lombard Laws*, transl. K. Fischer Drew, Philadelphia 1973, pp. 1–37; eadem, *Introduction*, [in:] *The Laws of Salian Franks*, ed. K. Fischer Drew, Philadelphia 1991, pp. 13–51; eadem, *Law and Society Early Medieval Europe*, London 1988.

25 A. Pertille, *Storia del diritto italiano dalla caduta dell'Imperio Romano alla codificazione*, vol. I–IV, Padova 1873–1887; P. del Giudice, *Le tracce di diritto romano nelli leggi longobardi*, [in:] idem, *Studi di storia e diritto*, Mailand 1889; N. Tamasia, *Le fonti dell'Editto di Rothari*, Pisa 1889; idem, *Storia del diritto italiano. Storia delle fonti dall'età romana ai tempi nostri*, Padova 1928; C. Giardina, *L'Editto di Rothari e la codificazione di Giustiniano*, Mailand 1937.

26 B. Paradisi, *Il prologo e l'epilogo dell'editto di Rothari*, Studia e documenta historiae et iuris, 34 (1968), pp. 1–31; G. P. Bognetti, *L'Editto Rothari come espediente politico di una monarchia barbarica*, [in:] *L'età longobarda*, IV, Milano 1968, 115–135; idem, *Fragmenti di uno studio Sulla composizione dell'Editto di Rothari*, [in:] *L'età longobarda*, IV, Milano 1968, pp. 585–609.

27 P. Delogu, *I Longobardi e la scrittura*, [in:] *Studi storici in onore di B. Bertolini*, I, Pisa 1972. idem, *L'Editto di Rothari e la società del VII secolo*, [in:] *Visigoti e Longobardi*, red. J. Arce, P. Delogu, Firenze 2001, pp. 330, 343; S. Gasparri, *La cultura tradizionale dei Longobardi, Struttura tribale e resistenza pagane*, Spoleto 1983; C. Azarra, *L'Italia dei barbari*, Bologna 2002.

in nature.[28] They are particularly interested in the relationship of the legislation of the Visigoth kings to the Roman law of those days[29], and also the problem of the territorial and personal nature of the law. Much attention has been devoted to the *Codex Euricianus* and the manuscript tradition of *Lex Visigothorum*.[30]

* * *

The research area of the human body, starting from the 1930s, has been the subject of many academic studies in the fields of history, anthropology and the sociology of culture. Certain currents in this research and its significance for the development of reflections on the history of the body in the Middle Ages have been presented in summary by Jacques Le Goff and Nicolas Truong.[31] The authors discuss the research of Marcel Mauss on the technique of using the body, the study by Norbert Elias into the process of adopting 'civilised' standards in the area of body control as well as in the physiological sphere, the works of Marc Bloch on the history of the body and the language of gestures connected with miraculous healings in royal rituals, Michel Foucault's investigation into the inclusion of the body in the microphysics of authority forms and its direct immersion in the sphere of politics.[32] The *History of the Body in the Middle Ages*

---

28   R. de Ureña y Smenjaud, *La legislación gótico-hispana*, Madrid 1905; J. M. Préndes Muñoz, *Historia de la legislación visigótica*, [in:] *San Doctor Hispaniae*, Seville 2002, pp. 50–67.
29   P. C. Díaz, R. G. Salirno, *El Codigo de Eurico y el derecho romano vulgar*, [in:] *Visigoti e Longobardi*, ed. J. Arce, P. Delogu, Firenze 2001, pp. 93–115; A. d'Ors, *La teritorialidad del derecho en la epoca Visigoda*, Annuaria de Historia del Derecho Espagñol, 5 (1959).
30   A. d'Ors, *El Codigo de Eurico. Estudios Visigoticos II*, Cuadernos del Instituto Juridico Espagñol, Rome–Madrid 1960; R. Gibert, *Códico de Leovigildo I–V*, Granada 1968. C. Díaz y Díaz, *La Lex Visigothorum y sus manuscripos. Un essayo de reinterpretación*, ADHE, 46 (1976).
31   J. Le Goff, N. Truong, *Une históire du corps au Moyen Âge*, Paris 2003, pp. 15-34.
32   Cf. M. Mauss, *Le technique du corps*, Journal de psychologie, XXXII, 3-4 (1936) in: *Sociologie et anthropologie*, Paris 1950; N. Elias, *The Civilising Proces. The History of Manners*, transl. by E. Japhcott, New York 1978; 1st German edition: *Über Process der Zivilization*, t. I-II, Basel 1939; M. Bloch, *Les Rois thaumaturges. Études sur les caractère surnaturel attribué á la puissance royale particulièrement en France et en Angleterre*, Paris 1961; English edition: *The royal touch: sacred manarchy and scrofula in France and England*, transl. by J. E. Anderson, London 1973; E. Kantorowicz, *The King's Two Bodies. A Study in Medieval Political Theology*, Princeton 1957; M. Foucault, *Surveiller et punir*, Paris 1975; English edition: *Discipline and punish; the birth of prison*, transl. by A. Sheridan, London 1977; idem, *Histoire de la sexualité*, vol. I-III, Paris 1976-1984; English edition: *The history of sexuality*, transl. by R. Hurley, New York 1978-1986.

is, however, first and foremost a synthetic depiction of many problems connected with the body and corporality located within a widely understood cultural notion, and particularly that of images, convictions and ideas. The authors' attention has been directed to such questions as: sexuality and procreation, diseases and treating the body, death and the dead body, ways of feeding, nakedness and costume, beauty and care over appearance, the metaphorical meanings of the body. This book also constitutes an attempt to sum up the oldest and most recent ascertainments on these questions.[33]

In the course of the last quarter century studies into the human body have become a fashionable area for research and have been undertaken many times in various disciplines of the humanities and social sciences (including medievalism). We shall mention here only a few of them. Within historical research into this problem attention is drawn to the works of Peter Brown on the reflections of early medieval clergy and philosophers on sexuality, by Caroline W. Bynum on the role of relics in the Christian Middle Ages, Jean-Claude Schmitt's study devoted to gestures in medieval culture.[34]

One needs to recall also the research into burial rituals and relations towards the corpse within European culture from late antiquity to the modern day.[35] Another research current in which the theme of the body and corporality plays an important role concerns the medical aspect of the problem. The question of health and illness as well as hygiene and diet has been analysed by G. Viga-

---

33 The work contains an extensive bibliography of the body and corporality broadly understood; this covers chiefly French works.
34 P. Brown, *The Body and Society. Men, Women and Sexual Renunciation in Early Christianity*, London-Boston 1988; C. W. Bynum, *The female body and religious practice in the Later Middle Ages*, [in:] *Fragments for a History of the Human Body*, ed. M. Feher, R. Naddaff, N. Tazi, New York 1989, vol. 1, pp. 161–219; idem, *Fragmentation and Redemption: Essays on Gender and the Human Body in Medieval Religion*, New York 1991; J.-Cl. Schmitt, *La raison de gestes dans l'Occident médiéval*, Paris 1990; cf. also: A. Paravacini Bagliani, *Il corpo del papa*, Torino 1994; idem, *The corpse in the Middle Ages: the problem of the division of the body*, [in:] *The Medieval World*, ed. P. Linehan, J. Nelson, London–NewYork 2001, pp. 327–341; cf. also: *Fragments for an History of the Body*, ed. M. Feher, vol. 1–3, New York 1989, particularly Jacques LeGoff, *Head or Heart? The Political Use of Body Metaphors in the Middle Ages*, vol. 3, pp. 13–26; *Framing Medieval Bodies*, ed. S. Kay, M. Rubin, Manchester 1994, particularly the article: M. Camille, *The image and the self: unwriting late medieval bodies*, pp. 62–99.
35 Cf. Ph. Ariès, *L'Homme devant la mort*, Paris 1977; M. Vovelle, *La mort en Occident de 1300 á nos jours*, Paris 1983.

rello.[36] In recent years there have appeared new studies into the role of the body in the Middle Ages.[37]

Research Methods into Barbarian Laws

The complexity of the researched area induces reflection as to the methods of interpretation for the source texts, and also into the body as a cultural and social category. As a result I would like to present the fundamental problems and methodological assumption undertaken in the research on the *leges* as textual statements relating to a so-called historical reality, as well as the human body. The legal codes of Germanic peoples will be treated first and foremost as an account speaking of the means of perception on the part of each of the tribal communities examined (although one formulated by a narrow group of experts) of their own specific social and legal reality. For the texts of barbarian laws may be interpreted as a specific discourse of their creators (in the broad understanding), which expressed *legal ideas, values and convictions* on society, the individual, social order and justice, the forms of law violation and the consequences for crimes committed.

The research conducted here will concentrate chiefly therefore on a reconstruction of the area of consciousness (views, convictions and intentions) of Germanic societies as contained within the texts of their laws. The aim of analyses thus presented is not, however, a reconstruction of the social and legal reality of the early medieval ethnic communities and tribal states composed out of them. For the subject of our research into barbarian legal codes is what one may call *the reality of text*. Here in this term I understand a specific vision and also a specific means of understanding and presenting historical reality. This approach is the result of a belief that the relationship between text and past reality *should not* be treated in the case of the source material herein under consideration as a simple direct reflection or equally registration through it of the social structures and situations of the researched epoch. In a similar way to narrative sources, equally in legal codes we are dealing with an image of reality presented through the medium of text. The key question is, therefore, the research into the content

---

36 G. Vigarello, *Le Sain et le malsain. Santé et mieux-être depuis le Moyen Âges*, Paris 1993; idem, *Le Propre et le sale. L'hygiène du coprs depuis le Moyen Âge*, Paris 1985; English edition: *Concepts of cleanliness: changing attitudes in France since the Middle Ages*, transl. by J. Birrell, Cambrigde 1988.

37 J. Le Goff, N. Truong, *op. cit.*; R. Mills, *Suspended Animation. Pain, Pleasure, and Punishment in Medieval Culture*, London 2005; B. Bildhauer, *Medieval Blood*, Cardiff 2006.

of the *leges* and the textual forms through which this is expressed, and therefore the structures of transfer themselves.

An important methodological problem of interest to us is also the specific pragmatics of the legal texts. For other types of source testimony such as historiographic, hagiographic or literary texts, which will be utilised in our analyses as comparative and contrastive materials, represent a vision specific to themselves as well as also versions of historical reality. Each of these types of transfer expresses, however, only the said vision in a separate way, in accordance with its own pragmatics. In legal texts, i.e., normative ones such as the *leges barbarorum*, it manifests itself in the fact that their basic if not only function was the regulating of the social life of a given people. The said regulatory function of the legal codifications of Germanic peoples or more broadly barbarian peoples has a fundamental significance for the way in which they are interpreted. For the law does not speak of that which has happened but about that which should have happened and more often what should not take place within social relations, and therefore about the proper and improper ways of human and group action as well as about the means for the resolution of disputes and problems.

Although much points to the fact that the majority of the research codifications were for a long time not applied in court practice, this in no way means, however, that barbarian laws were detached from actual social reality.[38] The legal questions contained within the individual *leges* regulations were strictly connected with the said 'social reality' of the early medieval Germanic kingdoms. It seems that each regulation presented, although in a model fashion and most selectively, a concrete situation known to the codifiers from an experience that potentially could have occurred in everyday life. The said social reality constituted, however, only one of the elements that comprised the construction which was to be the given legal rule. It did not appear within the regulations of the *leges* as such, but was transformed thus so as the regulations expressed the view of the creators of the law on the evaluation of a concrete situation and way of conduct in the case of a given crime being committed.

It follows to examine in the considerations of the research methodology for investigations into *leges barbarorum* the question of the specific concept and way of functioning of the law in societies like those of the early medieval Germanic peoples, representing a specific stage of cultural development. In other words, the matters concerns what the law was in traditional tribal societies, superficially Christianised, in which there dominated the culture of the spoken

---

38   H. Nehlsen, *Zur Aktualität*..., passim; C. Schott, *Pactus, Lex*..., passim; P. Wormald, *Lex Scripta*..., passim.

word. The research conducted by historians involved in the laws of early medieval peoples - Germanic, Celtic or Slavonic, out of necessity involves, first and foremost, analysis of the codification texts. The method of law research as conducted through analysis of normative texts supplemented by limited, particularly for the period prior to the year 800, accounts of a historiographic, diplomatic and hagiographic character talking of the practical application of the law, although valuable and indispensable, has for all that its limitations. For it leads to an identification of the law as the subject of research with its writing down, with the individual regulations contained in the barbarian *leges*. The concept of the law in traditional societies had, however, a wider scope than merely a collection of rules in writing regulating various specified legal situations.[39]

A Review of the Questions Examined in the Work

In undertaking research into the legal codes of Germanic peoples one needs to be aware of the limitations to the repertoire of subjects, connected with the body and corporality, that the said texts may be used for. This results from the specific character of legal texts. The matters regulated by the legal regulations that were contained in the texts of the researched codifications determine to a large degree a catalogue of problems which are to be the subject of our investigations. The human body in the barbarian *leges* is first and foremost the object of various types of crime involving the damage to its substance and violation to the bodily inviolability of man. This also appears in the context of crimes of a sexual nature both with the use of violence (particularly rape) as well as those which contravened social norms (e.g., prostitution, adultery). Other situations regulated by the law, where we have dealing with the body, are crimes against fertility and procreation. The next question is that of the body as the object of punishment – particularly mutilation as forms of punishment.

The main research material which we shall examine in the present work are Germanic legal codes understood as a specific form of textual utterance. An object of investigation thus defined implies the adoption of appropriate methods for the analysis of the source material. For it is first and foremost the language of these texts as an expression of the thought structures of the society that created them. We understand the category of the language for the researched texts here in a broad way. It covers both individual words (including specific terms

---

39 The existence and functioning of orally conveyed laws is known of, already after the writing down of the *leges* of the particular peoples, and these even in the case of those codifications which were to be used in court practice; for the Lombards *cawarfidae*; in the case of the Salian Franks – *lex salica non scripta*.

and formulas), sentences contained in the individual regulations, sets of regulations devoted to specific legal problems, but also whole legal codes understood as their own form of textual composition.

The language analysis of the sources will concern, although to varying degrees, all of these elements. We will be researching therefore the meaning of individual words and sentences, and also the structure of sets of regulations concerning crimes against the human body. These investigations will also take into consideration their wider context, which is the whole structure of the individual legal codes. An equally significant method of analysis will be the comparison of the cited language elements as contained in the particular legal texts. In our research we will make use of the extensive academic output that is the philological ascertainment of those historians who have engaged themselves in studies into the texts of barbarian laws. Here I am thinking first and foremost of research into the mallberg glosses contained in many of the codes and the language of the Anglo-Saxon codes.[40]

* * *

The problem area covered by the research involves several questions which will be discussed in the subsequent chapters of the book. In its first part (chapters 2 and 3) we will involve ourselves with the question of the individual's body. An important element in these investigations are the textual forms of presenting the body in the *leges* and in other medieval texts. This type of research approach requires also a comparison of legal texts with other types of historical documentation in which there appear the subject of the body and corporality. For here the matter concerns the presentation of the ways of comprehending concepts of the body as they appear in the *leges* within the cultural landscape of the early Middle Ages.

The first question I intend to research is the significance and way in which the concept of the human body was utilised in the legal texts of the Germanic peoples. This task will be served by comparative analyses of the individual codes.

---

40 H. Kern, *Notes on the Frankish words in the Lex Salica*, [in:] *Lex Salica. The Ten Texts within the Glosses and Lex Emendata*, ed. J. H. Hessels, London 1880, col. 431–564; F. Liebermann, *Gesetz der Anglesachsen*, vol. II, 1. *Wörterbuch*, Halle am Saale 1906 [reprint: Aalen 1960]; vol. II, 2. *Rechts- und Sachglossen*, Halle am Saale 1912; W. L. Van Helten, *Zu dem malbergischen Glossen un den salfrankischen formeln und lehnwörten in der Lex Salica*, PPB vol. 25 (1900), pp. 225–542; M. Gysseling, *De germanse woorden in de Lex Salica*, Verslagen en Mededelingen van den Koninklijke Academie voor Nederlandse Taal- en Letterkunde, Jg. 76, Gent 1976, pp. 60–109; R. Schmidt-Wiegand, *Die malbergische Glossen der Lex Salica als Denkmal des Westfrankischen*, Rheinische Vierteljahrblätter, vol. 33 (1969), pp. 396–422.

The investigated *leges* contain a catalogue of various forms of damage to body parts. The fundamental problem, which will be dealt with in Chapter 2, is the answer to the question: how did the authors of the codes see the relation between the individual body parts and the concept of the body as a whole? This matter is connected with research into the arrangements of the regulation sets that talk of damage and bodily violation contained in the various barbarian codifications.

The subject of the investigations will be also the factors shaping the means of presenting the body in the *leges* texts. These include: the subject of interest and the regulatory function of the legal texts; the culture of the spoken word and the written word as well as the types of thinking and knowledge transfer associated with it; the 'world outlook' resulting from Christian and pagan religious beliefs. The context in which the category of the body appears in the barbarian *leges*, is not merely the regulatory function but also the social phenomenon that was constituted by physical violence. Hence in the section devoted to the concept of the body are to be found analyses of descriptions of violence contained within other sources – historiographic, hagiographic and epic. Of interest for us will be especially the way in which the body was presented within the context of the said acts.

The next question, which we shall address in Chapter 3, is the mechanisms of bestowing value to the particular parts of the body. This analysis will concentrate on the reconstruction of the process by which the various systems of compensation amounts for various forms of bodily damage came into being. They will deal to a lesser degree with the amounts of fines for individual types of damage and bodily violation, while to a greater degree with what the mutual relationship was amongst the concrete material forms of compensation, and first and foremost their relationship to the entire structure of the compensation tariff system. The aim of this research is the reconstruction of those phenomena which may be defined as the social evaluation of the value of the losses incurred by a person as a result of damage to particular parts of the body. As a result the factors that decided upon the accrediting of a specific material value to definite forms of damage and violation were subjected to analysis. We shall involve ourselves in the following questions: the 'natural' hierarchy of body parts, the varied types of damage and violation as well as the circumstances surrounding the commitment of the said crimes.

In the further part of the work we shall examine the social aspect of crimes against the human body. Chapter 4 is devoted to the relation between the social differentiation of the Germanic peoples and the difference in the value of compensation awarded for the specific crimes against the body envisaged for groups of differing socio-legal status – freemen, freedmen and slaves. The main subject of the research is therefore in what way the groups of regulations that talk of crimes against the body of a representative of the said status groups were connected with the textual notions of proper or postulated social order. Another question that results from this is the relationship between the textual ideas of the body and social order. Besides we shall involve ourselves in a comparison in the approach adopted by the legislators in relation to the human body on the one hand and animal bodies (domesticated and wild) on the other. This problem is connected with the role of the human body (both living as dead) in the conception of man of the time as well as with the definition of society's framework and the determinants of affiliation to the human world.

Chapter One

The Law Books of Germanic Peoples as a Source for Research into the Human Body

We shall begin the presentation of barbarian codifications as source material from the assertion that several factors exist which define their contents (here equally those on the subject of the human body), which should be taken into consideration within our analyses. These include: elements of barbarian legal tradition, the specifics of cultural transfer within a culture based on the spoken word, the interaction of models taken from Roman legislation (including the idea of writing laws down) as well as the influence of the Latin (Roman-Christian) culture of the written word on the creation of written registers of barbarian laws. In order to define the means and scope within which the above phenomena formed the contents included in the rules of barbarian laws we should examine the origin of the *leges* texts, and first and foremost the causes of the origins and functions they were to perform in the understanding of their creators. Secondly, it is necessary to relate to the problem of the relations between the writing down of law and its oral transfer. Here the matter especially concerns the extent to which written law registered the custom norms orally formulated and transferred, as equally the influence of the process of registering and editing the *leges* texts in relation to their hitherto content. Of importance here is the defining of the chief features of the source texts researched in order to properly understand them and to explain their account on the human body as an object of criminality.

The Origin, Character and Functions of the Written Laws of the Germanic Peoples

An important task from the point of view of the characteristics of Germanic peoples serving as the source material for our research is the presentation of the most important theses that have been formulated to date in the extensive literature on the subject, on the mechanisms and historical circumstances of their arising, as well as their character and function. It is not our intention to conduct a systematic review or summary of the entirety of this research, we shall limit ourselves to references to selected, more recent publications which contain within themselves critical overviews of the thoughts of the earlier subject literature. We shall refer first and foremost to the works of P. Wormald and W. Sellert, and also H. Nehlsen, C. Schott.[41]

Wolfgang Sellert in the article *Aufzeichnung des Recht und Gesetz* (1992), which constituted an attempt at summarising the research conducted from the mid 19th century into the origin of the *leges* of the Germanic peoples, raises a fundamental question: for what reasons and with what aim were these said laws written down? This researcher approaches the matter with the assertion that „ergibt sich die zentrale und bisher nicht befriedigend gelöste Frage nach dem Sinn und Zweck der Stammes*auzeichnungen*"[42] ['the chief and to date unsatisfactorily solved problem is to illustrate the meaning and aims of encoding tribal laws'].

He subsequently considers the various explanations of this question. Before we present researchers' theories and views representing an attempt at resolving these problems in relation to individual law codes, we shall draw attention to propositions of a wider and more general comprehension of *leges* as a group of texts and their classification with regard to origin and functions.

Patrick Wormald in the article *'Lex Scripta' and 'Verbum Regis': Legislation and Germanic Kingship, from Euric to Knut* (1977) presented a series of important research theses, concepts and premises on the nature of Germanic legislation. Key here is the researcher's claim that Germanic law cannot be treated as a uniform group, particularly in relation to the functions ascribed it by its creators.[43] Developing this idea the author adopted the differentiation in force into the laws of peoples from the northern and southern parts of western Europe. On the one hand we have the codes of the Visigoths, Burgundians and Lombards, on the other the Franks, Anglo-Saxons, Alamanns, Saxons, Thuringians or Frisians. What constituted their distinctiveness was both the contents of the codes as well as the supposed, on the part of their creators, and actually fulfilled func-

---

41 P. Wormald, *'Lex Scripta'*..., passim; W. Sellert, *op. cit.*, passim; H. Nehlsen, *Zur Aktualität*..., passim; C. Schott, *Pactus, Lex*..., pp. 135–168; idem, *Zur Geltung*..., pp. 75–105.
42 W. Sellert, *op. cit.*, p. 69.
43 P. Wormald, *'Lex Scripta'*..., p. 107; the author equally claims that „ [...] it is much more difficult to generalize about early medieval legisation that many historians seems to have supposed. Partly, this is because of its variety, and partly because it yet obstinately refuses to fall neatly into clearly distinguishable categories.".

tions of the said law codes. Matters concerned here particularly the use or non use of written down law in court legal practice. In explaining the said difference the researcher writes:

> [...] *raison d'être* of the northern codes, involving a selection of custom, policy and judgment mixed up together, is far less obvious. Much barbarian legislation, in fact, gives the impression that its purpose was simply to get something into writing that *looked like* a written law-code, more or less regardless of its actual value to judges sitting in court.[44]

According to Wormald, the laws of the peoples of the northern zone were characterised by their selective character: 'Throughout northern Europe, – he writes – the issues selected for legislative record sometimes seem to have been dictated by arbitrary obsession rather than rational choice'.[45]

These views are not isolated. Herman Nehlsen, who undertook a detailed analysis of the legislation of individual Germanic peoples, here first and foremost into the legal texts of the Salic Franks in relation to the possibilities of applying their concrete regulations in court activities or other legal matters, came to the conclusion that "die *lex salica scripta* in der Rechtpraxis keine nennenswerte Rolle gespielt" [*lex salica scripta* did not play a greater role in legal practice].[46] Similar assertions have resulted from the research of Clausdeiter Schott into the encoding of Alamann laws.[47] While W. Sellert, referring to the ascertainments of historians doubting the practical application of the laws of the Salic Franks, Alamanns and other 'northern' Germanic peoples, claims:

> Die Konsequenzen daraus für die Frage der Rechtsqualität der Stammesrechte wären weitreichend. Da diese weder für die Praxis gelten, noch darauf zielen sollten, irgendwelche konkreten rechtlichen Verhältnisse wirksam zu regeln, könnte man fortan schwerlich überhaupt noch als *Rechte* und schon gar nicht als *Gesetze*, sondern bestenfalls als Scheingesetzgebung bezeichnen.[48]

---

44   *Ibidem*, p. 115.
45   *Ibidem*, p. 114.
46   H. Nehlsen, *Zur Aktualität...*, p. 449 ff.; cf. W. Sellert, *op. cit.*, p. 72: „Demgemäß führten nahezu alle in zeitgenössischen Urkundenmaterial enthaltenen Verweise auf die *lex salica scripta* ins Leere". ['almost all the indicators of *lex salica scripta* contained in contemporaray diplomatic sources lead nowhere']
47   C. Schott, *Zur Geltung...*, p. 75 ff.; idem, *Pactus, Lex...*, pp. 162–167.
48   W. Sellert, *op. cit.*, p. 76; in recounting the views of Schott the cited researcher writes: „Im Verhältnis zur mündlichen Rechtspraxis könnte sich daher das schriftliche Recht «auf einer phantastisch-unwirklichen Ebene» bewegt haben." ['Therefore in relation to the oral legal practice written law could have found itself "on the fantastic-unreal plane"'.] Cf. C. Schott, *Zur Geltung...*, p. 99. Sellert (*ibidem*, p. 74) examines the status of law written down in a historical context: „Die Germanen wiederum hielten ihr Analphabetentum für eine Tugend oder machten jedenfalls eine solche daraus. (...) Ingesamt

[The consequences resulting from this for the question of the quality of tribal law were far reaching. In as far as it did not have a practical significance or subsequently did not have as its aim the regulating of any concrete legal relations whatsoever, then it could not have had any significance from this point onwards either as a law or even more the case as an act but at best it could have functioned as pseudo-legislation].

Codifications in the southern zone had, according to Wormald, a different character. 'One can even rationalize – claims the author – the primary legislation of Visigoths and Lombards, as an attempt to convert their people to the use of *lex scripta*, by codifying basic custom on most issues and amending it where appropriate'.[49]

Southern Europe already during the period of the Early Middle Ages was to have been 'a pays du droit écrit (...) at least to a far greater extent then the areas to the north. (...) In Italy, Spain and southern Gaul, written *acta* remained central to legal procedure. (...) It was thus the environment that made the crucial difference to the scope and character of legislation in the southern Germanic kingdoms'.[50] Similar views on the subject of the functions of laws in the southern kingdoms of the Germanic peoples are presented by other researchers. H. Nehlsen claims that *Lex Visigothorum* was 'amazingly effective [...] in legal practice', while its regulations were used in the appropriate documents with the exact quotes as a legal basis.[51] W. Sellert points out in turn that Burgundian law 'not only envisaged the use of written documents for evidential aims' but also gave the costs of drawing them up.[52] This researcher also claims that there is a high degree of probability that Lombard laws equally played a significant practical role.[53]

The differences in the codifications of German peoples was derived, according to P. Wormald, not only from the variations in zones of civilization but also as a result of the cultural and political structural evolution of their kingdoms in the epoch under consideration. This is particularly visible in the case of the

---

dürfte bei den germanischen Völkern auch noch lange nach der Begegnung mit der spätantiken Kulturwelt das Analphabetentum vorherrschend gewesen sein. Der Laienadel und das Laienvolk lebten also schriftlos. ['The Germanic peoples [...] considered their illiteracy a virtue or at least made it out to be so. [...] Generally illiteracy might have maintained its grip over the Germanic peoples long still after the encounter with the world of late ancient culture. For the secular nobility and the secular people lived without letters.'].

49 P. Wormald, '*Lex Scripta*'..., p. 114.
50 *Ibidem*, p. 124.
51 H. Nehlsen, *Lex Visigothorum*, [in:] *HRG*, vol. II, col. 1966; cf. W. Sellert, *op. cit.*, p. 79.
52 W. Sellert, *ibidem*, cf. footnote 73.
53 W. Sellert, *op. cit.*, footnote 70.

Franks, for whom the function of written law in the Merovingian and Carolingian period was different. Therefore the author adopted a three way classification of the Germanic *leges*.

In those areas where the use of *lex scripta* was not only ordained but made easy, *lex scripta* was indeed used. In the areas where we find similar ambitions, but more marginal assistance to the judge, there are signs of move in this direction, but no more. In the parts of Europe where both instructions on, and manuscripts of, the law was are rare, there is scarcely a trace of the use of written texts in actual cases.[54]

The second fundamental question touched on by Wormald is the aim of creating written laws. This problem has also been considered by W. Sellert, who discusses at length the ascertainment and theses of historians (chiefly German and British), but also formulates his own conclusions, which differ from Wormald's position.[55] The British researcher started his exposition from a description of the contents of prologues to certain Germanic codifications, which proclaimed that the aim of a written publication of the law was the need of '(...) the promotion of peace and order, the redress of injustices and resolution of difficult cases (...)'.[56] However, he doubted in the value of argumentation based on these sources, claiming that it is difficult to explain the creation of all Germanic codes within the category of the need to introduce a just order. As a result of such an observation he proposed: 'We need to be able to postulate a *raison d'être* for barbarian legislation which takes account of all its features, including some which looks very little warts'.[57] He also presented his own views on the subject of the causes of the coming into being and functions of *leges*:

> (...) Germanic kings made laws, first and foremost, partly in order to emulate the literary legal culture of the Roman and Judeao-Christian civilisation to which they were heirs, and partly in order to reinforce the links that bound the king or dynasty to their people.

And also:

> (...) written law itself seems to have been inspired by ideological or symbolic considerations even for kings who certainly had strictly practical and legal objectives, and especially for the authors of primary legislation in northern Europe, it is also *direct* evidence for the image which Germanic kings and their advisers, Roman or

---

54 P. Wormald, '*Lex Scripta*'..., p. 122.
55 It is worth making a point here that Sellert, who otherwise cites the works of British researchers, almost completely passes over Wormald's theses; on the existence of the article quoted here he mentions the said, but once, in passing, in footnote 51 when commenting on the work of Wallace-Hadrill.
56 P. Wormald, '*Lex Scripta*'..., p. 105.
57 *Ibidem*, p. 106.

clerical, wished to project of themselves and their people: an image that might be immediate and political, or abstract and ideological; the image of king and people as heirs to the Roman Emperors, as counterparts to the Children of Israel, or as bound together in respect for the traditions of the tribal past.[58]

A different approach to the questions herein examined is presented by W. Sellert. Although he does not either totally reject the views of a part of the researchers (J. Wallace-Hadrill, C. Schott, G. Dilcher) that the laws of the Franks, Anglo-Saxons and Alamanns were created first and foremost for 'propaganda' or prestige aims and fulfilled rather symbolic rather than practical functions[59], however he claims that "[es] müssen daher über rein symbolische Zwecke hinaus noch andere Gründe für die Aufzeichnungen von Bedeutung gewesen sein." ['besides a meaning purely spiritual other reasons for their inscribing must have been of significance'].[60]

This supposition is substantiated by Rosamond McKitterick through ascertainments concerning the functioning of legal texts and in general the written word in the Carolingian epoch. For this British researcher claims that one cannot refer to the culture of the Frankish monarch for the period from the 8th to the 9th

---

58 *Ibidem*, p. 136.
59 J. Wallace-Hadrill, *Early Germanic Kingship in England and on the Continent*, Oxford 1971, pp. 37, 44: they were chiefly the subject of prestige and demonstration; C. Schott, *Zur Geltung...*, p. 101: they are understandable only as "Ausdruck fränkischer Herrschaftspotenz, sind Programm und Integrationssymbol des fränkischen Großreiches" ['an expression of the might of Frankish rule (authority) [...] the programme and symbol of the integration of the power of the Franks']. Sellert (*ibidem*, p. 78) on the basis of the ascertainments of G. Dilcher (*Gezetzgebung als Rechtserneurung*, [in:] *Festschrift für A. Erler zum 70. Geburstag*, ed. H. J. Becker, G. Dilcher, G. Guadian, E. Kaufmann, W. Sellert, Aalen 1976, p. 22 claims: „So liegt es nahe, daß die germanischen Könige mit einer Rechtsaufzeichnung ihre Stellung als christliche Herrscher und Inhaber der Staatsgewalt nach römisch-byzantinischem Vorbild herauszuheben beabsichtigten.". ['For it is possible that the Germanic kings through the encoding of the law intended to emphasize their position as Christian rulers and holders of stately power on the Roman-Byzantine model']; J. Wallace-Hadrill, *Early Germanic Kingship in England and on the Continent*, Oxford 1971, pp. 37, 44: they were chiefly the subject of prestige and demonstration; C. Schott, *Zur Geltung...*, p. 101: they are understandable only as 'an expression of the might of Frankish rule (authority) [...] the programme and symbol of the integration of the Frankish Empire'; G. Dilcher, *Gesetzgebung als Rechtserneurung*, [in:] *Festschrift für A. Erler zum 70. Geburstag*, ed. H. J. Becker, G. Dilcher, G.Guadian, E. Kaufmann, W. Sellert, Aalen 1976, p. 22: 'For it is possible that the Germanic kings through the encoding of the law intended to emphasize their position as Christian rulers and holders of stately power on the Roman-Byzantine model'.
60 W. Sellert, *op. cit.*, p. 80.

century as illiterate.[61] Sellert (in a similar way to McKitterick) doubts that almost 100 manuscripts of Salic law, coming from the period up until the end of the 8th century, did not have a practical application.[62]

In searching for an answer to the question as to the reasons for the writing down of the law the cited researcher quotes the views of H. Krause, according to whom the deciding motif was everywhere the 'resistance and durability' of its text.[63] One is inclined, however, towards the thesis of Gerhard Dilcher, that in this activity very often the matter concerned the renewal of a law.[64] Another motif, according to Sellert, for the creation of the said texts was the intention „die rechtlichen Beziehungen zwischen der unterworfnen römischer Bevölkerung und ihren Eroberern zu regeln" ['to regulate legal relations between the conquered Roman population and their conquerors'].[65]

> Ein eindrucksvolles Beispiel – argues the author – ist in diesem Zusammenhang die Lex Burgundionum. Obwohl auch dieses Stammesrecht nach der Einschätzung H. Nehlsens für die Praxis keine entscheidende Bedeutung haben sollte, paßt es sich doch bewußt ohne besondere Rücksicht auf das überkommene Recht, der veränder-

---

61 R. McKitterick, *Carolingians...*, Cambridge 1989, p. 1.
62 W. Sellert, p. 79; cf. R. McKitterick, *Carolingians...*, p. 56: 'It is difficult to believe that so many volumes of Germanic law were compiled in the Carolingian period for display purpose only". and further on: 'Undoubtedly they possessed a symbolic value, both as books per se, and as written law, but (...) they cannot categorically be denied a practical value as well.'
63 H. Krause, *Aufzeichnung des Recht*, [in:] *HRG*, vol. I Berlin 1971, col. 257. Sellert explains this idea in the following way (*ibidem*, p. 80): "«Bewahrung und Dauer» könnten bedeuten, daß mit der Aufzeichnung wie eine Beweisurkunde der Bestand und die Unumstößlichkeit des Rechts für alle Zukunft gesichert werden sollte;". [' "Resistance and durability" may mean that through inscription as with documentary evidence there was to be ensured the durability and irrevocability of the law in the future.']
64 G. Dilcher, *op. cit.*, p. 18; according to this author the monarch was someone more than merely an 'upholder of the mighty tribal legal tradition' rather 'the initiator of the writing down of the law and the necessary reforms to tribal law' (*ibidem*, pp. 25, 27).
65 W. Sellert, *op. cit.*, p. 85, further the author writes „Die Stämme haben dabei unterschiedliche Lösungswege eingeschlagen Teils haben sie die Rechte beider Bevölkerungsgruppen getrennt geregelt, teils sollten die germanischen Stammesrechte für beide Bevölkerungsgruppe verbindlich sein, wobei für die Römer, wo es notwendig schien, nur hier und da die eine oder andere Sonderregelungen getroffen wurde. ['Tribes chose here different solutions. In part they regulated the laws of both populace groups separately, in part Germanic tribal law was binding, while for the Romans, when it seemed necessary, only here and there were there undertaken one or another separate sets of regulations.']

ten Verhältnissen an und versucht die neu entstandenen Probleme durch entsprechende Vorschriften zu bewältigen.⁶⁶

[A forceful example – argues the author – is in this context Lex Burgundionum. Although this tribal law, according to H. Nehlsen's evaluation, also did not have a particular practical application, it follows to admit, however, that it adapts itself without particular consideration of the traditional law to changed relations and endeavours to solve anew the problems that have arisen through the appropriate regulations]

Sellert further considers how to explain the sense of the writing down of laws within a context, one claimed by H. Nehlsen and C. Schott, of the effectiveness of certain *legum*. More broadly the matter here concerns the relationship between the content of the text of law and the content of its oral version. He proposes the following explanation:

> Hier muß man sich zunächst von der Vorstellung lösen, die Rechtsanwender des frühen Mittelalters sein irgendeiner Subsumtionstechnik oder hermeneutischen Arbeitsweise verpflichtet gewesen. (...) Mit an Sicherheit grenzender Wahrscheinlichkeit dürfte es daher bei der „Rechtsfindung" kaum auf den genauen Wortlaut, sondern auf die allgemeinen Aussagen und Wertinhalte der Rechtstexte angekommen sein. Entsprechend dürften diese Texte in erster Linie „Bewußtseins- und Sinngehalte" vermittelt und zur „Institutionalisierung gemeinsamer Grundhaltungen" beigetragen haben.⁶⁷

[One needs to free oneself first and foremost from the notion that the user of law in the early Middle Ages was subjected to an technique of subsumption or a hermeneutic method of work. [...] Therefore with a likelihood verging on certainty one may state that in the 'finding of a law' the matter rather concerned not the exact wording but rather the general significance and value of the legal text's contents. Actually these texts could first of all present 'the sense and idea of the content' and result in the 'institutionalisation of a common principle position']

---

66 *Ibidem*, pp. 85–86; cf. H. Nehlsen, *Lex Burgundionum*, [in:] *HRG*, vol. II, col. 1901.

67 *Ibidem*, p. 87; the author here cites the idea of Th. Fögen (*Gesetze und Gesetzgebung in Byzantz, Versuch einer Funktionsanalyse*, [in:] IUS Commune, vol. XIV, hrsg. v. D. Simon, Frankfurt am Main 1987, p. 147); Sellert also substantiates his viewpoint with such an argumentation (*ibidem*): „Damit ließe sich (...) erklären, warum dort wo die zeitgenössischen Rechtanwender die Lex Salica zitieren, keine konkreten Bestimmungen dieses Stammesrecht auszumachen sind. (...) Liegt es doch jetzt nahe, daß sie jedenfalls auch als Grundlage für eine *mündliche* Verkündung des jeweiligen Rechtstexten am Königshof, in den Grafschafts- sowie den Heers- und Thingversammlungen in Betracht kamen. ['With this [...] one may explain why where the contemporary users quoted *Lex Salica*, there cannot be recognised any concrete regulations from this legal code. [...] It is now, however, close to the truth that they [i.e. written down laws] in any case also entered into the equation as the basis for every "oral" announcement of every legal text at the royal court, at earl, lord and council gatherings.'].

The Law Books of Germanic Peoples as a Source for Research into the Human Body 37

He also negotiates the question as to the function of the law into a thesis on its practical role in court activities. He refers here to the ascertainments and views of Ruth Schmitt-Wiegand on the *leges of the* Salic Franks and Alamanns:

> An die Einführung der *lex scripta* wurden hohe Erwartungen geknüpft. Wie sehr man dabei die Praxis in Auge hatte, zeigen nicht zuletzt die Glossierungen der Lex Salica und der Lex Alamannorum, die eine im einzelnen allerdings noch immer nicht überzeugend geklärte „Verbindung zwischen dem aufgezeichneten Recht und der Rede vor Gericht herstellen" und insoweit den beabsichtigen „schrittweisen Proceß der Verschriftlichung des Rechts in der Zeit vom 6.-8. Jahrhundert" belegen.[68]

> [Great expectations were connected with the implementation of *lex scripta*. How much practice is taken here into consideration is shown by, not at all in the smallest degree, the glosses in *Lex Salica* and in *Lex Alamannorum*, which create a still not entirely convincing detailed explanation of 'the link between written law and rhetoric before a court' though to a degree they confirm the intended 'gradual process of writing down of the law from the 6th to the 8th century'].

On the other hand, he sets forth the view that „(...) der mündliche Befehl, das *verbum regis*, wie ein Gesetz wirksam werden und das bisherige Recht aufheben oder verändern." ['(...) the oral order, *verbum regis*, could also have been effective as an act to repeal or change the existing law.'].[69] He confirms this with Hanna Vollrath's thesis, who, in researching Anglo-Saxon laws in relation to the role of the spoken and written word, claimed that literacy and the law did not need to create a whole, for in Anglo-Saxon England the concept of *lex* as *constitutio scripta* had still to arise.[70] Besides the cited researcher considers:

---

68 Ibidem, p. 90; R. Schmitt-Wiegand, *Malbergische Glossen*, [in:] *HRG*, vol. III, col. 211–212; eadem, *Franken und Alamannen*, [in:] *Festschrift für Karl Schmid*, ed. K. Althoff, Sigmaringen 1988, p. 69; Sellert (*ibidem*, p. 92) in citing G. Radbruch's idea (*Einführung in die Rechtswissenschaft*, ed. K. Zweigert, Stuttgart 1964, p. 13) he writes also „(...) das Recht einzig und allein durch seine schriftliche Fixierung eine derartige Heraushebung erfahren haben, daß es alsbald konkurrierend neben die Gewohnheit «als zweite jüngere Rechtsquelle» trat, aus der *in nuce* nicht mehr «das Herkommen, sondern [der] menschliche Wille sprach», dem es nun freistand, «das überkommene mündliche Recht» zu billigen oder zu verwerfen." ['(...) law only in its written registration enjoyed this type of recognition so that shortly in contention it was to become, next to custom, "the second younger source of law", from which *in nuce* now no longer "custom but the will of the people spoke", thanks to which it was possible to accept or reject "the law conveyed by oral tradition" '.]

69 W. Sellert, *ibidem*, s. 92.

70 H. Vollrath, *Gesetzgebung und Schriflischkeit. Das Beispiel der angelsächsischen Gesetze*, Historisches Jahrbuch, vol. 99 (1979), p. 54; In commenting on this notion (*ibidem*, s. 93 footnote 148): „Schon die Tatsche, daß die Angelsachsen ihre Stammesrechte nicht in Latein, sonder in ihrer eigenen Sprache aufgezeichnet haben, spricht dafür, daß

(...) hinreichend nachgewiesen worden ist, im frühen Mittelalter nicht zwischen *lex scripta* und *lex non scripta* unterscheiden. Lex konnte daher auch mündliches Recht bedeuten, was damit zusammenhängen könnte, daß die aufgezeichnete *lex* neben Sitte und Gewohnheit undifferenziert als gleichwertig (...)[71]

[(...) it has been sufficiently well shown that in the early Middle Ages *lex scripta* and *lex non scripta* were not differentiated between. Therefore *lex* may represent also a spoken law, which could have been connected with the fact that written laws were perceived next to custom and habit as being of equal worth (...)].

Finally though he sides with another viewpoint. On the basis of the accounts from *Lex Baiwariorum*, the laws of Liutprand, the Frankish capitularies and *Lex Gundobada*, Sellert concludes: „Ingesamt zeigt sich in Belegstellen die Tendenz, den schriftlichen Recht den Vorrang vor der Rechtsgewohnheit einzuräumen." ['generally in the evidence there appears a tendency to ascribe precedence to written law in relation to legal custom.'].[72] Sellert connects the act of law recording and the functioning of *leges* first and foremost with the intention for a strengthening of royal authority. This was to come about through subordination

---

ihnen ein Gesetzbegriff, wie ihn Isidor rezipiert hatte, fremd war. ['Already the very fact that the Anglo Saxons wrote down their laws not in Latin but in their own language tells one that for them a conceptualisation of the law as it was seen by Isidor (of Seville), was alien'] – *Lex est constitutio scripta; mos est vetustate probata consuetudo sive lex non scripta*.

71 *Ibidem*, p. 95. The author makes reference to the works of Theuerkauf, Kroeschell, Köbler, Weitzel and others; G. Theurkauf, *Lex, Speculum, Compendium iuris. Rechtsaufzeichnung und Rechtsbewußtsein in Norddeutschland vom 8. Bis zum 16. Jahrhundert*, Köln–Graz 1968, p. 60; K. Kroeschell, *Recht und Rechtsbegriff im 12. Jahrhundert*, [in:] *Probleme des 12. Jahrhunderts*, Konstanz– Stuttgart 1968, pp. 309–319; G. Köbler, *Das Recht im frühen Mittelalter*, Köln–Wien 1971; J. Weitzel, *Dinggenossenschaft und Recht* (Quellen und Forschungen zur höchsten Gerichtsbarkeit im Alten Reich, ed. B. Diestelkamp, U. Eisenhardt, D. Gudian, A. Laufs und W. Sellert, vol. 15/II Köln–Wien 1985, p. 1357; following the idea of J. Weitzel (*Deutsches Recht*, [in:] *Lexicon des Mittelalters*, vol. 3, München–Zürich 1986, col. 779) Sellert (*ibidem*) claims: „(...) könnte man annehmen, daß die ersten Rechtsaufzeichnungen »in ihrem Wesen und ihren Wirkungen zunächst und grundsätzlich durch die Vorstellung vom mündlich geübten Überzeugungsrecht bestimmt« waren." ['(...) one may accept that the first written laws were "defined first and foremost in their essence and effect and fundamentally by the image of an orally applied legal conviction" ']

72 *Ibidem*, pp. 98–99; cf. footnotes 170–171, 174–179; the author claims „Die Ausschließlichkeit der *lex scripta* kommt auch dort klar zum Ausdruck, wo der König sich vorbehält, das schriftliche Recht zu ergänzen." ['the exclusivity of *lex scripta* takes its turn where the king has conditioned things in order to supplement written law.']

of the law and legislation to the monarch.[73] Hence in his summarising of the whole exposition Sellert claims:

> Die Aufzeichnung der Stammesrechte gehört zu den revolutionierenden Ereignissen der deutschen Rechtsgeschichte. Durch sie sollten die Grundlagen für eine mit dem Willen des Königs geschaffene, auf Dauer angelegte, verbindliche und steuerbare Rechtsordnung gelegt werden. Damit geriet die überlieferte und mündliche bestimmte Rechtskultur ins Wanken. Nun wurde die Frage, welches Recht denn gelten sollte, virulent und zugunsten der *lex scripta* entscheiden. Zwar brauchen Aufzeichnung und Gesetz nicht unauflöslich zusammenzuhören und sich gegenseitig zu bedingen. Indem aber dem schriftlichen Recht der Vorrang eingeräumt und *per definitionem* nur noch dieses als *lex* gelten sollte, wurden die entscheidenden Weichen dafür gestellt, daß von nun an die Rechtsgewohnheit an die zweite Stelle rücken sollte.[74]

[The writing down of tribal laws is one of the most revolutionary occurrences in German legal history. Thanks to this there were to be lain the bases for the accomplishment with the king's agreement of a permanent established, binding and regulated legal order. Through this the traditional orally transferred legal culture started to falter. Hence the question as to which law had significance was solved to the advantage of *lex scripta*. Admittedly the record and law did not require the creation of an indissoluble whole and were not mutually conditioned to each other, while precedence was given to written law and *per definitionem* it already was as *lex* to have meaning. The decisive points were so fixed that from then onwards custom law was to be pushed into second place.]

The above conclusion raises doubts. Sellert, so it seems, is examining the matter of writing down the *leges* from the point of view of the historical effects of the move from traditional law conveyed orally to a specified law, recorded and created in a written form on the order of the authorities. This premise prejudges to a

---

73 *Ibidem*, p. 99: Mit der Absicht, der *lex scripta* die alleinige Geltungskraft zu verschaffen, wurde zugleich eine Konsolidierung der königlichen Herrschaft bezweckt. Denn »das Setzen und Durchsetzen von Normen ist ein wesentliches Element von Herrschaft«. Dem König sollte dementsprechend nach dem Vorbild des christlich-byzantinischen Herrschers die ausschließliche Rechtssetzungmacht und Interpretationsherrschaft über die *lex* zustehen. Der König sollte »als zentrale Gewalt die einzige Instanz« sein, die Einheit des Stammesrecht in dem Zeitpunkt lebendig zu erhalten, »als es sich nicht mehr als mündliche Tradition zu reproduzieren« vermochte. ['The intention for only *lex scripta* to obtain validity was intended to strengthen royal authority. For the "establishment and forcing through of norms is an important element of rule." The king was to be vested appropriately to this, according to the Christian-Byzantine model of a ruler, the ruler determining the law and the interpretation of *lex*. The king was to be "like the central authority in a single instance" who at a given moment held the unity of tribal law, "for it was not possible to be reproduced longer as an oral tradition"']; (quotes cf. H. Vollrath, *op. cit.*, p. 584; Wieacker, *op. cit.*, p. 35; G. Dilcher, *op. cit.*, 25).

74 *Ibidem*, p. 102.

large degree these very conclusions. If we examine this phenomenon from the perspective of around the year 900 (i.e., at the decline of the epoch of barbarian laws), such an evaluation has indeed justification; yet if one were to take as the reference point the start of this process i.e., the time when the first laws of subsequent peoples were written down (from the start of the 6th to the beginning of the 9th century), the status of particular codifications appears varied. At this stage within the culture of the majority of the peoples herein described oral communication dominated – including the oral transfer of law; court and administrative procedures were conducted orally, there was a strong participation of old 'tribal' laws. This state of affairs was considered the natural, proper one. In studying the epoch between circa 500 and circa 800 it follows to take into consideration this starting point.

In summing up it follows to state that both the research premises as equally the ascertainments of Wormald, Nehlsen and Schott that have been reviewed above are cognitively valuable and inspirational. For these researchers have shown the features of the legal texts researched by them, which existed, during the period of their creation and later as well, as the cultural context in which they functioned, not being influenced by the effects of the cultural changes which had started the recording of them in writing.

The Problem of the Relationship between the Recording of Law and the Oral Legal Tradition

In considering the key question for an evaluation of the cognitive value of *leges barbarorum* and their character as source evidence, we should relate to the problem of the relationship between the content of the said records and the oral legal tradition of the Germanic peoples of the early Middle Ages. In other words how the collections of regulations written down in individual barbarian laws related to the oral version of law that had functioned before, as well as also after its capture in a written, textual form.[75] The explanation of this problem is important from the point of view of the research herein undertaken though at the same time difficult (in as far as it is possible at all), for the oral transfer of law is unavailable to us. This does not mean, however, that we are completely devoid of possibilities to research the question.

75　Here the matter does not concern the relations between *lex scripta* and *lex non scripta*, particularly in the context of the problem of applying written law in court practice or administrative matters, but about the content of the law contained in *leges*.

Another key problem of the subject under consideration is the answer to the question: whether and to what degree the preserved texts of barbarian codification (written down law) can be treated as a reflection of oral legal transfer, and also as an expression of the oral mentality that created the legal tradition of the Germanic peoples of the early Middle Ages. Here, however, arise subsequent questions: Did the said oral transfer initially contain a traditional and systematic structure bestowed in the past by then legal specialists? If it possessed the said then of what kind? And also – what happened to it during the process of text editing: was it rejected or accepted in its entirety or was it also modified? If it did not possess a definite structure – then in what way (according to which criteria and at which stage of the codification process) did this hypothetically unordered collection of rules have bestowed on it a compositional structure? These questions are particularly important in relation to the groups of rules on crimes against the body that are contained in the laws under study.[76]

The difference between law conveyed orally and that written down is connected to a large degree with the division introduced by older researchers of the *leges* – into custom or folk law – on the one hand and royal or statute law on the other.[77] From our point of view this distinction is significant so much so that custom laws arose and functioned initially in the spoken language and only later received a written form, while royal law was formulated from the very outset in

---

[76] Cf. J. Goody, *The Logic of Writing and the Organization of Society*, Cambrigde 1986. The author formulates the following view (p. 133): „[...] the encapsulating of oral practices as written rules has far-reaching consequences for the members of a society. The written code does not initiate either oppression or justice; it gives them a different format, whiche relates to the modes of communication and is not merely a matter of changing one set of cultural clothes for another." In another place (*ibidem*, p. 129) the author writes: „As suggested in a much earlier article, by creating a text 'out there', a material object detachet from man (who created and interprets it), the written word can become the subject of a new kind of critical attention. [...] Moreover the text is often more difficult to understand since its lacks the context of speech, may well be abbreviated, cryptic and generalized [...] Moreover the criterion of legal text involves a formalization (e. g. a numbering of the laws), a universalization (e. g. an extention of their range by the elimination of particularities) and ongoing rationalization."; cf. J. Goody, I. Watt, *The consequences of literacy*, [in:] *Literacy in Traditional Society*, ed. J. Goody, Cambridge 1968, pp. 27–84.

[77] Cf. J. Goody, *The Logic...*, pp. 129–130, writes: 'In Europe the distinction between law and custom is ultimately based on what was written an what was not. To codify is to set it down in writing before proclaiming it as law'. And further: 'In the jural systems of societies without writing, there can be no effective distinction of this kind between law and custom'.

writing.[78] The relations between the oral and written mode of law creation is a complex problem – for in practice the rules initially formulated in writing could display certain features of the oral utterance, while on the other hand the norms of 'pure' oral origin would have undergone in the process of notation certain modifications resulting from the regulations of the literary system. However, it seems that this genetic difference between 'custom' law and 'statute' law, understood as a certain general tendency, is in our case a useful instrument for the description and characteristics of the *leges*.

In a recently published work on the oral tradition of the Icelandic sagas, Gísli Sigurðsson considers the problem of research into the oral sources of the oldest law of the Icelandic people - *Grágás* (written down in the period 1117–1118). He raises questions important for the matters here under consideration: „How it is possible to investigate this oral tradition? And what is the relationship between this tradition and the written texts we find in the manuscripts of the Middle Age and later?".[79] In writing about the earlier stage of development for the culture of the country he claims: „Before the arrival of the Church, Iceland was an oral society. People learned stories and poems one from another and practiced their religion and governed themselves with laws and legal judgments without the aid of writing. (...) The training of legal experts was still entirely oral as late as the early years of the 12th century.".[80] This, in a sense, fairly obvious viewpoint is important, for it characterises the situation of the majority of peoples herein dealt with during the period of the initial editing of their *leges*.

* * *

Many often divergent and sometimes conflicting views on the subject of the appearance within *leges* texts of the traditional legal norms that had arisen (and been formulated) prior to them being written down exist within the subject literature. It is neither possible nor necessary to present here the entirety of these viewpoints and pieces of research. They have been and are still referred to, discussed and subject to critical evaluation by contemporary medieval scholars.[81] It

---

78 Cf. Wormald, *Inter cetera*..., p. 977.
79 G. Sigurðsson, *The Medieval Icelandic Saga and Oral Tradition. A Discourse on Method*, transl. N. Jones, Cambridge, Mass.– London 2004, p. 53.
80 *Ibidem*, pp. 55 and 57.
81 P. Wormald, *Lex Scripta*'..., passim; idem, *The Making*..., passim; H. Nehlsen, *Zur Aktualität*..., passim; W. Sellert, *Aufzeichnung*...; C. Collins, *Law*...; T. M. Charles-Edwards, *Law in the western kingdoms between the fifth and the seventh century*, [in:] *The Cambridge Ancient History*, vol. XIV, *Late Antiquity: Empire and Successors, A. D. 425–600*, ed. A. Cameron, B.Ward-Perkins, M. Whitby, Cambridge 2000, pp. 260–285; K. Modzelewski, *Barbarzyńska Europa*, Warszawa 2004; French edition: *L'Europe*

follows, however, to refer to some of the more important theses, in particular the positions developed during the course of the last thirty years.

The prominent representative of the German school of the history of law (*Rechtsschule*), Heinrich Brunner, on the basis of research into *Lex Salica* and other laws remaining within its sphere of influence, has formulated the view that barbarian law constitutes an expression of *Volksgeist*.[82] Researchers forming the said school have also differentiated two types of legislation functioning within the Frankish monarchy – *Lex* and *Capitularia*. *Lex* was to have been in their opinion folk or tribal law in nature, while *Capitularia* was to be royal law.[83] P. Wormald, basing himself on the research of F. L. Ganshof, considers, however, the likely division of Frankish legislation was into: originally what was to have been '(...) the original statement of a people's law in writing" as well as additionally '(...) concerned either to amend, reinforce or supplement the first statement, or with only administrative matters'.[84] P. Wormald, referring to the laws of the Germanic peoples, has claimed, although not without reservations, that '(...) there are indirect indications that much of what we find in Germanic codes does represent the custom of the relevant people as it was conceived at the time (...)'. As an example of the said customs the author points to the systems of *wergelds* and compensation, which is considered to be one of the oldest tribal elements of the laws of barbarian peoples.[85]

---

des barbares. Germains et Slaves face aux heritiers de Rome, transl. by A. Kozak, I. Macor-Filarska, Paris 2006; German edition: *Das barbarische Europa: zur sozialen Ordnung von Germanen und Slawen im frühen Mittelalter*, transl. H. Petersen, Osnabrück 2011; idem, *Legem ipsam vetare non possumus. Królewski kodyfikator wobec potęgi zwyczaju*, [in:] *Historia, idee, polityka. Księga ofiarowana J. Baszkiewiczowi*, Warszawa 1995, pp. 26–32.

82   H. Brunner, *Deutsche Rechtsgeschichte*, vol. I, p. 406.
83   H. Brunner, *op. cit.*, pp. 406–412, 418–420, 540–551; cf also P. Wormald, *'Lex Scripta'...*, p. 109
84   F. L. Ganshof, *Was waren die Capitularien?*, transl. W. Eckhardt, Weimar 1961; P. Wormald, *'Lex Scripta'...*, p. 110.
85   *Ibidem*, pp. 111–112; Recently P. Wormald (*The 'Leges Barbarorum': law and ethnicity in the post-roman West*, [in:] *Regna and Gentes. The Relatioship between Late Anique and Early Medieval Peoples and Kingdoms in the transformation of the Roman World*, eds. H.-W. Goetz, J. Jarnut, and W. Pohl, Leiden-Boston 2003, p. 41) proposed an interesting thesis: 'The compensation tables of the "Ripuarians", as of Alamanns and Bavarians, do in fact differ in detail from those of the "Salic" Franks (see Appendix below). To contemplate the broadly similar yet invariably divergent "tariffs" of the "*Volksrechte*" raises the thought that these tables were in some sense *the* differing factor in any one *lex*, expert knowledge of which thus identified the legal specialists. [...] A tariff recognizable as one's own was perhaps itself an ethnic marker.'

Historians researching the question of the recording of the laws of Germanic peoples have formulated various views on the relationship between the recording of these laws and legal tradition. In relation to certain codifications the majority of researchers accept that only to a limited degree did they reflect the tribal unwritten laws of the Germanic peoples. This especially concerns the notations of the law that came into being in the 5th and beginning of the 6th century. In German literature and earlier Spanish there is acknowledged the participation of Germanic legal concepts in the creation of the oldest Visigothic codification i.e., *Codex Euricianus*.[86] While A. d'Ors has interpreted the legislative activities of Euric as a continuation of the *ius edicendi* of the Roman prefects.[87] *Codex Euricianus* is in his opinion 'a relic of Roman common law'. The Germanic elements such as the cleansing oath and the compensatory system, was to have been, according to the researcher, in essence merely primitivisms and manifestations of decadence, next to which classic Roman law started once again to develop.[88] H. Nehlsen presents a more balanced thesis that „[i]n zahlreichen Fällen kann der Codex Euricianus als Zeugnis genuin westgotischen beziehungsweise (...) durchaus auch germanischen Rechtsvorstellungen betrachtet werden", and also that „(...) für jüngere Phasen der westgotischer Gesetzgebung ist eine Anlehnung an das vulgarisierte römisches Recht zu beobachten. In nicht geringem Umfang sind jedoch auch hier noch Rechtsvorstellungen zu erkennen, der Ursprung im germanischen Kulturkreis gesucht werden darf.[89] ['in numerous cases *Codex Euricianus* may be viewed as original Visigothic evidence as a result [...] also of its completely Germanic legal concepts', and also that '[...] for the younger phases of Visigoth law making one may observe a basis upon common Roman law. However, to a substantial extent there are recognisable legal concepts here whose sources may be searched for in Germanic cultural circles.'].

A most unequivocal standpoint, similar to that of d'Ors, is expressed in turn by Roger Collins, who writes about *Codex Euricianus*: '(...) the earliest Visigothic one does not consist of a writing down of traditional Germanic laws or customs. The text that has been identified as the only surviving section of the Code of Euric is predominantly, even exclusively, Roman in its contents and its structure'.[90] According to him the rules of this code constituted a significant part

---

86 H. Nehlsen, *Lex Visigothorum*, [in:] *HRG*, vol. II, col. 1974; The author cites the position held by B. R. Ureña, *La legislacion gotico-hispana*, Madrid 1905.
87 A. d'Ors, *El Codigo*, p. 17.
88 *Ibidem*, pp. 2, 9.
89 H. Nehlsen, *Lex Visigothorum*, [in:] *HRG*, vol. II, col. 1975.
90 R. Collins, *Visigothic Spain...*, p. 227; cf. also, idem, *Law and Ethnic Identity...*, pp. 2 and 17, idem, *Early Medieval Spain. Unity in Diversity, 400–1000*, London 1995 [1983], pp. 27–30.

of the codification *Liber Iudiciorum*, preserved in its entirety, and issued by King Recceswint in 654. 'In addition to laws of Chindasuinth, and Reccesuinth, – writes Collins – the *Liber Iudiciorum* also included 315 other laws lacking royal attribution, but which are headed *Antiqua* (...)' coming presumably from the end of the 5th century.[91]

In discussing the Burgundian *Liber Constitutionum* H. Nehlsen claims that in a different way than the Lombard *Edictum Rothari*, it was not "(...) als umfassende Aufzeichnung der *antiquas leges patrum ... que scriptae non erant* angelegt war. Vielmehr handelt es sich um eine Sammlung legislativer Maßnahmen aus besonderen Anlässen."[92] ['(...) planned as an extensive notation of *antiquas leges patrum* [...] *quae scriptae non erat*. The matter rather concerned a collection of legislative rulings created for specific reasons.']. According to the researcher even rules formulated on the basis of older Germanic models contained new contents. This manifested itself in, among other things, a departure from the initial system of compensation and a transition to capital punishments and subsequently to the introduction or increase in public fines, which existed next to private forms of damages.[93]

The Burgundian *Liber Constitutionum*, according to R. Collins, displays many features of Roman common law.[94] 'It is reasonable to assume – he claims – that the code was compiled from the royal edicts of Gundobad and Sigismund, original examples of which are to be found in four *constitutiones extravagantes* [...]' incorporated into the text of the book of laws.[95] The legislative activity of the Burgundian kings was modelled on the Roman practice of creating collections of laws by provincial prefects. Collins also believes that this codification displays few features and elements of Burgundian tribal legal tradition.[96] A similar view is expressed by Katherine Fischer Drew, who has formulated the following thesis: „Thus, the *Lex Gundobada* represents a tend away from tribal custom based on moral sanction to royal enactments based on political authority and power of the king".[97]

---

91   R. Collins, *Visigothic...* p. 234; *Leges Visigothorum* [further: *LVisig*], MGH LNG, vol. I, ed. K. Zeumer, Hannover 1902.
92   H. Nehlsen, *Lex Burgundionum*, [in:] *HRG*, vol. II, col. 1909.
93   *Ibidem*.
94   R. Collins, *Law...*, p. 9; I. Wood, *The Code in Merovingian Gaul*, [in:] *The Theodosian Code*, ed. J. Harries, I. Wood, New York 1993, p. 174; edition *Liber Constitutionum* [further: *LConst*] in: *Leges Burgundionum*, MGH LNG, vol. II, 1, ed. L. R. von Salis, Hannover 1892.
95   R. Collins, *Law...*, p. 9.
96   *Ibidem*.
97   K. Fischer Drew, *Introduction* [in:] *The Burgundian Code...*, p. 10, she writes also: „In many respects the Burgundian legislation seems to be a body of tribal customs that had

R. Collins, in considering recently the matter of the character of the laws of Germanic peoples formulated in the 5th and 6th century (i.e., those of the Visigoths, the Ostrogoths, and the Burgundians) claimed that '(...) a number of the codes might become little more than random and arbitrary selections from the much greater mass of oral and customary law of the people, and in the consequences only partial and limited guides of the latter'.[98] This position results from the author's view that: '(...) the extant collections of laws issued in the western kingdoms in the fifth and sixth centuries mirror the character of prefectural edicts (...)', and also that during the period of lawmaking they presented continuity in relation to the legislative practice of the Western Empire.[99]

The recording of the law of the Salic Franks is a complex matter and one which has been explained in a number of ways by particular researchers. One set of theses has been advanced by Ruth Schmidt-Wiegand, making recourse to older literature. She claims that *Lex Salica* (and particularly its first edition, the so-called *Recensio Chlodovea*) „erhält sie auch kaum Spuren antiken Rechts" as well as that „[d]er Inhalt der Lex verrät vielmehr auf Schritt und Tritt fränkische Sicht." ['contain only a few traces of ancient law' (i.e. Roman) as well as that 'the content of *Lex* displays at every turn a Frankish viewpoint.'] In the formal aspect, according to her, Salic law was of the consensual legislative type (*Pactus*), which was formulated in the form of sentences (*Weistümer*) found by the proponents of the law.[100] She adds, however, after Franz Beyerle, that *Lex Salica* displays two different types of norm: the titles speaking of the compensation (*Buß*) and the constitutions.[101]

R. Schmidt-Wiegand also deals with the tale of the creation of Salic law notation as contained in the so-called 'Short Prologue'. Despite the legendary nature of the names of those who founded the laws (Wisogastus, Arogastus, Salegastus, Widogastus) Schmidt-Wiegand believes that „[e]inen historischen

---

evolved considerble distance in direction of positive statue law, especially in case of *novellae*. In its successive additions and modifiactions there can be seen a trend toward establishment of a body of royal legisation; that is, rather simple customary rules have evolved into a more complex royal legislation."

98 R. Collins, *Law...*, p. 2.
99 *Ibidem*, p. 17.
100 R. Schmidt-Wiegand, *Lex Salica*, [in:] *HRG*, vol. II, col. 1950. A similar view in relation to the subject of Frankish Salic law is expressed by Clausdieter Schott (*Der Stand...*, p. 36): 'Within the group of the oldest Germanic codifications *Lex Salica* distinguishes itself in form and content through its archaic character.' *Pactus legis Salicae* [further: *PLSal*], MGH LNG, vol. IV, 1, ed. K. A. Eckhardt, Hannover 1962; *Lex Salica* [further: *LSal*], MGH LNG, vol. IV, 2, ed. K. A. Eckhardt, Hannover 1969.
101 *Ibidem*, col. 1954: F. Beyerle, *Über Normtypen...*, pp. 220–225.

Kern wird man in dessen auch diesem Prolog nicht ganz absprechen können"[102] ['it would not be, however, possible to deny this Prologue a historical core.']. A similar viewpoint has been presented recently by Karol Modzelewski.[103] In researching the accounts on the mechanisms behind the process of noting in a written form the oral account of the law contained in the *leges* texts themselves (particularly: the 'Short Prologue' of *Pactus legis Salicae*, the laws of the Frisians, and also the 'Prologue' of *Lex Baiwariorum* and the 'Epilogue' of *Edictum Rothari*), the author has come to the conclusion that the activities undertaken for the purpose of writing down the laws started from the oral announcement on the part of legal specialists of legal sentences on all court matters.[104] The said sentences were noted down by a scribe but under the direction of legal experts such as the four mentioned in the 'Prologue' of Salic law or Wlemar and Saxmundus appearing in the record of Frisian law.[105]

A completely different position on Salic law has been recently proposed by Roger Collins. On the basis of a comparative analysis of the mechanisms by which the legislation of the Visigoths, Burgundians and Ostrogoths came into being as well as the nature of the said on the one hand, and that of the Salic Franks on the other, he has drawn into question the thesis that *Lex Salica* constitutes a codification of tribal legal customs. According to him, such a thesis '(...) gains no support from a survey of the other [Germanic] codes' of the 5th and 6th

---

102   *Ibidem*, col. 1951.
103   K. Modzelewski, *Barbarzyńska...*, pp. 54–56; the author writes: 'the oldest information about how matters proceeded during the codification of legal customs is contained in the so-called "Short Prologue" to the Salic law. This prologue is of the 6th century and in principle deserves belief despite the fact that it was added to the body of legal norms some time after their notation.' Referring chiefly to the law advocates named in the 'Prologue' he claims that 'Salegast and Widogast should be recognised as eponyms of the locations Saleheim and Widoheim situated to the east of the Rhine. The place names given in the prologue [...] clearly point to the places of origin of those selected for the codification. They were without doubt Franks from the uniform German areas of Austrasia.'
104   Cf. The Short Prologue in: *PLSal*, § 2: *Extiterunt igitur inter eos electi de pluribus uiri quattuor his nominibus: Uuisogastus, Arogastus, Salegastus Uuidogastus „in uillas quae ultra Rhenum sunt: Bothem, Salehem et Uuidohem", qui per tres mallos conuenientes omnes causarum origines sollicite discutientes de singulis decreuerunt hoc modo* [...]; and also *Additio Sapientum* [further: *Add. Sap.*] in: *Lex Frisionum* [further: *LFris*], MGH Fontes iuris Germanici antiqui XII, ed. K. A. Ekchardt, Hannover 1982, III after § 48: *Haec iudicia Wlemarus dictavit*; before § 59: *Haec iudicia Saxmundus dictavit*; Modzelewski in commenting on the account of the prologue *Pactus legis Salicae* claims: 'The issuing of a sentence at a gathering involved the announcement of a relevant norm of Frankish Salic law; that was *dicere leges salicam*" (*ibidem*, p. 55).
105   K. Modzelewski, *Barbarzyńska...*, p. 59.

century. In none of the above mentioned codifications (including the Salic) are there visible, in his opinion, an intention 'to produce a written statement of supposed ancestral Germanic legal customs'.[106] *Lex Salica* (the so-called Salic Law Pact), according to Collins, may be interpreted as 'a systematic collection of early Frankish royal edicts, removed from their original contexts and deprived of any reference to the legislators who issued them.'[107] Collins also questions the significance of the 'Short Prologue' of *Pactus legis Salice* as an argument in the thesis that it was a record of old Frankish law. This prologue was to have been – in his opinion – 'a part of the mythical history of the Franks', still alive in the second quarter of the 8th century; he also draws attention to the fact that the tale of the four men pondering all court matters during three consecutive assembles was added to *Liber Historiae Francorum*, whose author located the writing down of the law of the Salic Franks during the times of King Faramund, the grandson of Trojan Priam, which as he judged would be in the second half of the 4th century.[108]

In relating to the question of the origin of the regulations within *Pactus legis Salicae*, Ian Wood has adopted a somewhat different, less radical, view. He takes his line of argumentation from the claim that this law code did not contain the whole of Salic law. He also claims that 'Much of the law in the *Pactus legis Salicae* probably did equate with that known by the *rachinburgi* (...)', which means its oral, traditional version.[109] A part of the laws were, according to the researcher, of royal origin however. 'To some extent, therefore, the *Pactus Legis Salicae* and the *Lex Ribvaria* resemble the *Liber Constitutionum* of the Burgundians issued by Sigismund in 517, for without question this last text includes a number of royal edicts. In the Frankish codes such laws are included alongside other, plausibly customary, legislation.'[110] I. Wood also points out the non-Christian, and even pagan, character of some of the *Pactus* regulations, which agrees with the view as to the traditional, tribal origin of a part of the regulations contained in it.[111]

---

106 R. Collins, *Law*..., p. 14.
107 *Ibidem*, p. 15
108 *Ibidem*, p. 14
109 I. Wood, *The Merovingian Kingdoms, 450–751*, London–New York 1994, p. 110.
110 *Ibidem*. The author give examples of rules that were, in his opinion, royal edicts not derived from ordinary (common) law: a) those concerning the king's authority (footnote 66): XIV, 4; XVIII; XLI, 8; LVI, 1, 4, 5; b) those relating to Romans (footnote 67): XIV, 2–3; XVI, 5; XXXII, 3–4; XXXIX, 5; XLI, 8–10; XLII, 4; c) to slaves taken abroad (footnote 68): XXXIX, 2. Therefore in total only 9 from amongst 65 items in *Pactus*, in part or in whole, did not belong to the tribal legal tradition of the Salic Franks.
111 *Ibidem*, p. 113; as an example the author gives title II, 16, where talk is about an oxen given as sacrifice (footnote 76).

Remaining within the area of the Merovingian Kingdoms let us move to a description of the record of the law of the Ripuarian Franks – *Lex Ribuaria*.[112] In German studies there exists the viewpoint that this law code is not a record of tribal law in the traditional understanding but simply a subsequent form in the development of Salic law, created for the needs of the Ripuarian Franks. R. Schmidt-Wiegand points to the influences of Burgundian law and through its intermediacy certain elements of Roman law. „Die Lex Ribuaria – he writes – ist das einzige »Stammesrecht«, das die Konstitution Chlotars II. über die Gerichtskunde und damit modernstes reichsfränkisches Urkundenrecht rezipiert hat". [113] ['*Lex Ribuaria* – she writes – is the only 'tribal law' which adopted the constitution of Chlotar II on court documents and through this is the most modern document law of the Frankish kingdom.']. I. Wood, in turn, emphasises that *Lex Ribuaria* contains elements of legal tradition (tariffs for compensation for damages to the body), regulations derived from or modelled on the Salic *Pactus* as well as royal edicts.[114]

From the point of view of the question here under discussion the recent ascertainments concerning the oldest recordings of Anglo-Saxon law appear to be especially valuable, and particularly the law of Æthelberht, king of Kent. Patrick Wormald emphasised the archaic character of the oldest Anglo-Saxon codification, particularly at the language level. 'One of the most striking characteristics of Æthelberht's laws is the extreme simplicity of their syntax.'[115] He pointed out also other features of this legal code which emphasised its 'archaic' and also 'primitive' character: the lack of procedural elements and the weak influence of Christianity.[116]

Already earlier researchers into the laws of Æthelberht had accepted that they could constitute the record that was the closest to the oral transfer of law. Lisi Olivier in a recently published work on the beginnings of Anglo-Saxon law has advanced the following thesis: 'If we assume, based on the linguistic evidence (...), that the text of Æthelberht does in fact represent a record of archaic Anglo-Saxon laws, they are the first Germanic laws to be recorded in vernacu-

---

112 *Lex Ribuaria* [further: *LRib*], MGH LNG, vol. III, 2, ed. F. Beyerle, R. Buchner, Hannover 1954. cf. also col. 1923-1937
113 R. Schmidt-Wiegand, *Lex Ribuaria*, [in:] *HRG*, vol. II col. 1925.
114 I. Wood, *The Merovingian*, p. 115; idem, *Administration, law and culture in Merovingian Gaul*, [in:] *The Uses of Literacy in Early Medieval Europe*, ed. R. McKitterick, Cambridge 1990, p. 66.
115 P. Wormald, *'Inter cetera'*..., p. 969. In analysing the law in its entirety, Wormald claimed that 75% of the provisions are simple conditional sentences beginning from 'if' (*gif*), e.g., 'if one lies with a king's maiden, let him pay fifty shillings', *ibidem*, p. 970.
116 *Ibidem*.

lar. As such they stand boldly at the watershed between orality and literacy in the Anglo-Saxon legal tradition.'[117] Also valuable are the views on the subject of the structure of the discussed legal code made by Olivier:

> Excluding the first provisions dealing with the church and public assembly, these are presented in a top-to-bottom order. They deal first with the king, then move to his household, to his nobles, and finally turn to the freemen of land. Then follow the personal injury laws, and finally the laws regarding those whose status differs from that of freemen: women *esnas* or servants, slaves. (...) The personal injury laws illustrate a head-to-toe mnemonic similar to the top-to-bottom ordering of the text as a whole. This section consist of thirty-nine clauses (33-71) laid out in order from the top of head to the ripping of the big toenail.[118]

G. P. Bogneti's research into the composition of *Edictum Rothari* has shown the presence of a huge block of archaic legal sentences (*Weistümer*) containing tariffs for damages and next to them Rothari's intention to introduce a resilient royal organisation of courts and administration.[119] H. Nehlsen, who has analysed the *Edict's* regulations on slaves, claimed that '[...] Rothari's legislation re-

---

[117] L. Olivier, *Beginnings...*, p. 36. The author refers to the research of Dorothy Bethurum (*Stylistic Features of the Old English Laws*, The Modern Language Review, vol. XXVII (1932), pp. 263–279), who ascertained that '[t]he large number of legal formulas, many of them alliterative, many of them rhymed, which repeat themselves throughout the decrees of Old English and Old Frisian laws, the Sachsenspiegel, and the Old Norse laws, lend some probability to the hypothesis of an original poetic form of the Germanic laws.' p. 266. Also in other Indo-European languages, particularly in Old Icelandic, there exist analogical mnemonic poetic procedures serving the recording of legal tradition.

[118] L. Oliver, *Beginnings...*, p. 36: "'The overall structure of Ae's laws can be laid out as follows: 1-7 Offences against the church and public assembly, 8-17 Offences against the king and his households, 18-19 Offences against *eorlas*, 'noblemen', 20-32 Offences against *ceorlas*, 'freemen', 33-71 Personal injury laws, 72-77 Offences against (and rights of) women, 78-81 Offences against *esnas* (a rank with a legal status intermediate between freeman and slave), 81-83 Offences against *þeowas*, "slaves"'; see: *Æthelbert* [further: *Æthel*], [in:] *Die Gesetze der Angelsachsen*, ed. F. Liebermann, vol. 1., Halle 1891 [reprint: Aaalen 1960], pp. 5–7; cf. also translations into English: *The Laws of the Earliest English Kings*, ed. F. L. Attenborough, Cambridge 1922, pp. 5–17; and into German, *Leges Anglo-Saxonum 601–925*, ed. K. A. Eckhardt, Göttingen 1958; cf. also the works on Anglo-Saxon law: W. S. Simpson, *The Laws of Ethelbert*, [in:] *On the Laws and Customs of England. Essays in Honor of Samuel E. Thorne*, ed. M. S. Arnold, Th. A. Green, S. A. Scully, St. D. White, Chapell Hill 1981, pp. 3–17; P. Wormald, *The Making...*, passim; H. Vollrath, *Gesetzgebung und Schriftlichkeit...*, pp. 28–54.

[119] G. P. Bogneti, *Fragmenti...*, pp. 583 ff.; edition of the Edict in: *Leges Langobardorum*, MHG Leges, vol. IV, ed. F. Bluhme, Hannover 1868; *Le leggi die Longobardi*, ed. C. Azarra, S. Gaspari, Milano 1992, pp. 12–119.

mained, given the precision in language, under the influence of Roman models, however the content of the regulations is fairly independent [...]".[120]

Stefano Gasparri, who is examining the oldest Lombard codification within the context of the transfer from oral culture to the culture of writing, claims that: "Lo stesso Editto, nelle sua complessità, si rivela ben più di una semplice raccolta di una serie di norme penali. Esso raduna la somma di quelle consuetudini – cawarfidae è la parola langobarda latinizata – che tutte insieme constituivano le regole generali di vita della *gens Langobardorum*."[121] ['The *Edict* in its complexity turns out to be rather a simple collection of a series of penal norms. It collected a sum of legal customs – *cawarfidae*, i.e., the Latinised Lombard speech – which together created the general regulations of life the *gens Langobardorum*']. In another place he writes: "[editto] nella sua forma più anticha ed orale veniva sicuramente recitato in forma poetica: nella stessa redazione scritta vi sono di ciò tracce indubitabili, rappresentate da parole germaniche dal valore alitternte (*lid in laib, in gaida et gisil*)."[122] ['(the edict) in its oldest and oral form was for certain recited in a verse way: traces of it undoubtedly exist in the written edition itself, presenting Germanic speech of alliterative value (*lid in laib, in gaida et gisil*)'].

According to another Italian historian, Claudio Azarra, *Edictum Rothari* "propone dunque come la raccolta meditata di tutto patrimonio normativo consuetudinario, trasmesso oralmente, che viene per la prima volta fissato in un testo scritto.[123] ['presents itself as a well-thought out collection of the entire inheritance of custom norms, conveyed orally, and then for the first time recorded in a written textual form.'] According to him "la legislazione longobarda presenta

---

120 H. Nehlsen, *Sklavenrecht...*, pp. 358 ff., and particularly p. 412 ff.; cf. G. Dilcher, *op. cit.*, p. 25; B. Paradisi, *op. cit.*, pp. 1–31.
121 S. Gasparri, *La memoria storica dei Longobardi*, [in:] *Le leggi die Longobardi*, ed. C. Azzara, S. Gasparri, Milano 1992, p. VI. "La cultura longobarda, come tutte le culture dei popoli germanici, era dunque una cultura orale; ed è proprio nel momento del suo passagio dall'oralità alla scrittura, che, se da una parte si perde il dinamismo originario di quella cultura, destinata, una volta consegnata alle regole dello scritto, a mutare profondamente i suoi caratteri originari, dall'altra si creano le condizioni per una forma, sia pure parziale e distorta, di conservazione di contenuti antichissimi." ['Lombard culture was as in the case of all the cultures of Germanic peoples, an oral one; it is characteristic of the period of its transfer from orality to literacy that on the one hand this culture lost its initial dynamism, which at a certain moment is subjected to the regulations of writing, and is designated for a deep change in its original character', while on the other hand there arise conditions allowing for the preservation of the oldest texts though in a form that is only partial and distorted'].
122 *Ibidem*, p. XI.
123 C. Azarra, *Introduzzione al testo*, [in:] *Le Leggi...*, p. XXV.

infatti cartatteri di grande originalità, offerendo una materia genuinamente germanica nella sua essenza, assai più »pura« (...), di molte altre codificazioni barbari altomedioevali."[124] ['Lombard legislation crucially presents features of high originality, offering material that in its essence is truly Germanic, which is far 'purer' [...] than many other early medieval Germanic codifications.']. Relating to the philological aspect of *Edictum*, Azzara claims that it contains numerous elements of the former Lombard language in nature traditional formulas as well as alliterative clauses. Of fundamental significance to us is the view of the author that "(...) le leggi sono dapprima pensate in langobardo, in una forma alliterante e cantilene che è tipica dei modi della trasmissione orale, e vengono poi tradotte in latino da redattori romani (...)"[125] ['(...) laws were first formulated in Lombard, in an alliterative form, something that was typical for oral transfer, and were subsequently translated into Latin by Roman editors (...)'].

The law making of the early Alamanns is represented by two texts: *Pactus legis Alamannorum* and *Lex Alamannorum*.[126] K. A. Eckhardt has ascribed the older edition to the Frankish king Chlotar II (584–629), dating the time of its creation to the first decades of the 7th century. The creation of *Lex* he accredits to the Alamann prince, Lantfryd, while the time period during which the compilation arose he places within the period when he was loyal to the Frank monarch, i.e., the period between 712 and 725.[127] The initial arrangement of the *Pact of Alamann Law* is not clear. It is not to be excluded that initially it covered regulations concerned with the *wergeld* for murder and arrangements for compensation and bodily violations.[128] The Alamann *Pact* has archaic and pre-Christian features.[129] *Lex,* however, divides itself into three parts on: the Church, the prince and the people.[130] The third part of *Alamann Law* constitutes in general the old-

---

124 *Ibidem*, p. XXVI.
125 *Ibidem*. p. XXVII.
126 *Leges Alamannorum*, MGH LNG, vol. 5, 1, ed. K. Lehmann, Hannover 1888 [2nd edition by K. A. Eckhardt, Hannover 1966]; this edition contains both versions of the Alamann law: *Pactus legis Alamannorum* [further: *PLAl*] and *Lex Alamannorum* [further: *LAl*]; cf. also *Leges Alamannorum*, Germanenrechte, Neue Folge,Westgermanisches Recht, vol. 1, *Einführung und Recensio Chlotariana (Pactus)*, Göttingen–Berlin–Frankfurt 1958, vol. 2, *Recensio Lantfridiana (Lex)*, Witzenhausen 1962.
127 K. A. Eckhardt, *Einleitnung*, [in:] *Leges Alamannorum*, vol. I, p. 90, vol. II, p. 7; cf. C. Schott, *Der Stand...*, p. 40; idem, *Die Leges*, p. 15; cf. also: Th. Rivers, *Introduction*, [in:] *Laws of the Alamanns and Bavarians*, ed. Th. Rivers, Philadelphia 1977, p. 24, dates the *Lex* to the years 717–719.
128 C. Schott, *Die Leges...*, p. 17.
129 C. Schott, *Lex Alamannorum*, [in:] *HRG*, vol. II, col. 1885.
130 C. Schott, *Pactus...*, p. 141; similarly: Th. Rivers, *op. cit.*, p. 24, cf. *Lex Alamannorum*, regulations on: the Church – titles 1–22, the prince – 23–43, the people – 44–98.

est regulation materials of Alamann tribal law. They present the entire complex, containing a broad listing of legal rights and in the case of their violation information about compensation. In this manner there were evaluated more or less all living entities, from a free Alamann of average status, for whose death one had to pay 200 *solidi* right up to his yard dog the value of which was a *solidus*.[131] *Lex* clearly shows the features of strong Church interpolations and Church cooperation in establishing the law. The most important intention of these regulations may be seen in their defence of the Church and their permeation with Church matters.[132] Alamann law is relatively rich also in Frankish phrases and Franko-Latin and specifically Alamann ones.[133]

C. Schott, in relating himself to the relations between written law and the oral tradition in Alamann codifications, claims that „Eine Gesellschaft wie die alemannische des 7. und 8. Jahrhunderts lebt fast ausschließlich in einer schriftlose Ordnung, in der eingeübte und tradierte Normvorstellungen weithin vorherrschen. Es handelt sich um Gewohnheitsrecht im Sinne einer unreflektierten Rechtsordnung. Schriftrecht ist vor allem dann erforderlich, wenn das alte Recht Traditionsschwächen zeigt oder wenn es neuem Recht zur Durchsetzung verhelfen soll."[134] ['Alamann society in the 7th and 8th century lived almost exclusively in a state of illiteracy, in which learnt and traditional normative concepts greatly predominated. The matter here concerns custom law in the meaning of a non-reflective legal order. Law written down was needed first and foremost when the old norms displayed the weakness of traditions or when it was to aid in the forcing through of a new law.'].

A subsequent collection of barbarian laws is *Lex Baiwariorum*.[135] The time when this text was written down is unclear. Eckhardt and Buchner have assumed that it arose in the 740s, during the times of Prince Odylon i.e., before the year 748.[136] According to K. Beyerle this text was to be edited at the monastery of

---

131 C. Schott, *Die Leges Alamannorum*, [in:] *Lex Alamannorum. Das Gesetz der Alemannen, Text – Übersetzung – Kommentar zum Faksimile aus der Wandalgarius-Handschrift Codex Sangallensis 731*, Augsburg 1993, p. 19.
132 C. Schott, *Leges Alamannorum*, [in:] *HRG*, vol. II, col. 1884.
133 R. Schmidt-Wiegand (*Alemannisch und Fränkisch in Pactus und lex Alamannorum*, [in:] *Beiträge zum frühalemannischen Recht*, ed. C. Schott, Brühl–Baden 1978, p. 9 ff.) has shown the stages in the creation of the text and the lexical layers of this folk language in the particular case of *Codex Sangallensis 731*.
134 C. Schott, *Pactus...*, p. 162.
135 H. Siems, *Lex Baiwariorum*, [in:] vol. II, col. 1887–1901; *Lex Baiwariorum* [further: *LBaiw*], MGH LNG, vol. V, 2, ed. E. von Schwind, Hannover 1926.
136 C. Schott, *Der Stand*, p. 41.

Altaich shortly after its founding i.e., in the year 743/744.[137] The Bavarian codification shows the influences of Visigoth law. *Lex Baiwariorum* also displays many common features with *Lex Alamannorum*. In the Bavarian codification, in a similar way to that of *Lex Alamannorum*, there appears a threefold content arrangement, subsequently there were herein placed regulations on Church matters (title I), the prince (titles II-III) and the people (title IV-XXII). There also exist certain structural similarities to *Edictum Rothari* (the catalogue of compensations for crimes against the body is comprised of three parts, of which the first concerns freemen, the second those freed, and the third slaves).

According to H. Siems, the record of the law of the Bavarians did not arise as a completely planned text, but rather through the process of gradual development.[138] He also claims that defining the chronology of its coming into being and therefore the origin of individual rules is difficult „da alle Anschauungen in der Lex Baiwariorum mit jüngerem Nachträgen und auch Einarbeitungen altüberkommener Sätze des Gewohnheitsrechtes rechen"[139] ['because all the ideas contained in *Lex Baiwariorum* incorporate later additions as well as the inculcation of the former, traditional principles of custom law.']. This researcher raises the question as to „ob nicht die lange Bewahrung einzelner Rechtssätze durch eine frühere Redaktion erst ermögliche wurde"[140] ['whether the long term maintaining of individual legal norms was not made possible by their earlier editing?']. K. Modzelewski, in writing about the *Prologue* claims that although the information about the committing to paper of these *leges* by the Frankish ruler, Theodoric I (511–534), is untrue, the account contained within it of the means by which Bavarian law came about may be considered as illuminating. 'The author of the prologue – he claims – realised that a written codification had to maintain the old customs, which meant the involvement in the undertaking of experts of the oral legal tradition.'[141]

The legal codes of the Saxons, the Thuringians and the Frisians belong to the latest barbarian *leges* written down upon the initiative of Charlemagne, presumably at the gathering of secular and religious notables at Aachen for the years 802-803. Saxon law is represented first and foremost by *Lex Saxonum*. This notation is preceded, however, by two other texts of a legal nature: *Capitu-*

---

137 K. Beyerle, *Einleitung*, [in:] *Lex Baiwariorum*, hg. von K. Beyerle, München 1926, p. LXV.
138 H. Siems, *Lex Baiwariorum*, [in:] *HRG*, vol. II, col. 1897.
139 *Ibidem*, 1890.
140 *Ibidem*.
141 K. Modzelewski, *Barbarzyńska...*, p. 53.

latio de partibus Saxoniae (782 or 785) and *Capitulare Saxonicum* (797).[142] R. Schmidt-Wiegand has defined *Lex Saxonum* as the written compilation of Saxon tribal law. This results, in her opinion, from the different nature of this codification in relation to *Capitulatio*. For there do not appear in it regulations connected to the process of the Christianisation of this people, which were contained in *Capitulatio*. She also draws attention to the presence of many rules traditional in character (family law, property and inheritance, titles LX–LXIX). The characteristic feature of the Saxon codification, which distinguished it from the other barbarian legal codes, is the particular position which was ascribed to the magnate stratum (*nobiles*, *Edelinge*). This expressed itself first and foremost in the amount of *wergeld* and the compensation awarded for this group, which was six fold greater than that given to ordinary freemen (*ingenui*).[143]

The writing down of Thuringian law (*Lex Thuringorum*) has been, relative to other barbarian legal codes, poorly researched.[144] Johannes Herold, who published the code in 1557, defined it as *Lex Angliorum et Vuerinorum hoc est Thuringorum*. According to R. Schmidt-Wiegand, this legal code was, in character, a collection of court sentences (*Weistümer*). In relation to the construction and terminology employed it showed many similar traits in common with *Lex Saxonum* and *Lex Frisionum*, which presumably results from the fact that they came into being at the same time and in similar circumstances.

A special case with regard to the mechanism of creating its written version is constituted by the law of the Frisians. The text of the said code, which has come down to us through the intermediary of the print prepared by Johannes Herold in 1557, is comprised of two parts *Lex Frisionum* and *Additio sapientum* (headings presumably given by the publisher).[145] Researchers involved in Frisian law have come to the conclusion that the source issued by Herold was not the final edited version of the work of those who had taken the pains to write the

---

142 G. Theuerkauf, *op. cit.*, pp. 47–48; C. Schott, *Der Stand*, p. 42; *Capitulatio de partibus Saxoniae* and *Lex Saxonum* (further: *LSax*) [in:] *Leges Saxonum et Lex Thuringorum*, MGH Fontes iuris Germanici antiqui, ed. C. F. von Schwerin, Hannover 1918.

143 R. Schmidt-Wiegand, *Lex Saxonum*, [in:] *HRG*, vol. II, col. 1963; cf. also K. Modzelewski, *Barbarzyńska...*, pp. 234 ff.

144 Cf. R. Schmidt-Wiegand, *Lex Thuringorum*, [in:] *HRG*, vol. II, col. 1965–1966; C. Schott, *Der Stand...*, pp. 41–42; *Lex Thuringorum* (further: *LThur*), [in:] *Leges Saxonum et Lex Thuringorum*, MGH Fontes iuris Germanici antiqui, ed. C. von Schwerin, Hannover 1918.

145 H. Siems, *Lex Frisionum*, [in:] *HRG*, vol. II, col. 1917.

law down, but simply a working copy of the legal code.[146] As a result of which K. Modzelewski claims:

> Despite the name bestowed by Herold, *Additio sapientum* is connected with the initial link of codification works. This was neither an appendix nor a semi-product but mere raw material. The manuscript in which a sixteenth-century publisher saw a collection of court sentences was in point of fact an officially recorded notation of the words of two proponents of the law devised for the purpose of codification.[147]

The author also writes that *Additio* and *Lex* constitute a trace of the two-stage process of creating the text of Frisian law; An *Addition* was to have been the original notation of the oral version of the law, while *Lex* was to have been the version of the text that underwent the editorial and censorial processing.[148]

* * *

It is not our intention to comprehensively research the question of the 'literacy' or 'oral' origin of the content for all the regulations and rules or individual norms of behaviour within each of the law codes of interest to us. We will look only at those groups of rules that concern crimes committed against the body, and more strictly violation of its integrity and bodily inviolability. We shall examine the compositional arrangement of the said groups of rules as they appear within the individual barbarian *leges*. In particular we shall be interested in the means of grouping together, as well as the ordering, in which the individual case of criminality against the body are enumerated. The solutions that have been adopted in this matter within the researched legal codes will be viewed by us against the background of the entire compositional scheme that appears in each one of these.

The composition of the groups of regulations of interest to us correspond, in the case of the majority of the *leges*, to the structural entirety of the individual texts. While within the legal codes one may note a tendency to group together rules concerning similar or related legal matters (e.g., crimes against property, family law etc.), the regulations devoted to crimes against corporeality create within the legislative code structure a more or less separate uniform thematic block. While in those codes where individual legal questions were not presented according to an ordered scheme, rules regulating cases of bodily violation and harm do not create comprehensive groups.

Man's body as an object of criminality was presented in the individual records of laws by means of several different arrangements of the group of regula-

---

146 Cf. W. Krogmann, *Entstehungszeit und Eigenart der Lex Frisionum*, Philologia Frisica (1962), p. 76; C. Schott, *Der Stand*, p. 42; K. Modzelewski, *Barbarzyńska*, p. 57.
147 *Ibidem*, p. 58.
148 *Ibidem*, p. 56.

tions concerning violation and bodily harm. One of these, the most frequently met, was characterised by an attempt to depict these regulations in a more or less extended comprehensive class, constituting a separate section for a given legal code. This model is represented in its entirety by the *Law of Æthelberht*, *Lex Ribuaria*, *Edictum Rothari*, *Pactus legis Alamannorum* and *Lex Alamannorum*, *Lex Saxonum*, *Lex Frisionum*, with it functioning to a lesser degree in the laws of the Visigoths (i.e., in *Liber Iudiciorum*, from the mid 7th century and in *Antiqua* of the third quarter of the 6th century, and possibly also in *Codex Eurici*, at the latest from the beginning of the 6th century). This practice resulted from the fact that offences against an individual person were perceived as one of the main subjects of interest of the law amongst barbarian peoples (next to theft, murders, inheritance matters and family affairs). The principle for organizing the structure of these sections of the *leges* was the adoption of a scheme of viewing the human body from the head down to the feet. The second solution also involved the creation of a substantively united set of regulations. There was not applied within this, however, the said 'natural' arrangement of enumerating individual crimes based on the build of the body itself. This type is represented by the *Pactus* and the *Lex* of the Salic Franks, *Lex Baiwariorum* as well as *Lex Thuringorum*. In the third compositional variant the collection of researched rules did not create within the codification text a separate entirety. Individual regulations were noted down without any specified ordering scheme. The only legal code with such traits and features is the Burgundian *Liber Constitutionum*.

It follows to here emphasize that the meaning of the proposed classification is not reduced to a description of the formal features of the structure within the herein discussed groups of rules. It is intended to serve in reconstructing the ways of perceiving and presenting man's body within the individual barbarian *leges*. The analysed codifications constituted with regard to the textual forms a presentation of the body as an object of diverse offences, a varied collection. The conjecture is raised as to whether the outlined compositional systems could have constituted an expression of various means of perceiving the human body, one that arose within the context of various cultural systems.

The characteristics of the various *leges* presented above show that their record was the result of a complex game played out between influences Roman-Christian and Germanic in origin and encompassing the culture of the written word and oral accounts. Taking into consideration the hitherto ascertainments cited by the researchers above, concerning the mechanisms and circumstances of this writing down of individual legal codes, one may formulate the following hypothesis: in the specific *leges* we are dealing with a perception of the human body appropriate for Roman-Christian culture, for Germanic-pagan culture or for some formation combining certain elements of both of these.

Chapter Two

The Concept of the Human Body in the Germanic Legal Codes and in Early Medieval Narratives

It is possible and advisable to discern within the research into the human body as it manifests itself in the *leges* both an approach within which it is viewed as a phenomenon that is the subject of the imagination, and consequently something that is inter-subjective, as equally one in which it is treated as an 'objective' phenomenon. In the first of these approaches the matter concerns the reconstruction of the specific way in which the term was understood and used, and the grasping of its cultural meaning within the historical form that it was to take within the legal codes of the Germanic peoples of the early medieval period. The latter approach consists in the use of a contemporary concept of the body utilised within the confines of a dictionary understanding of the concept of the body when analysing the content of those *leges* that correspond to this 'objective' category. The concept of the body is formed through cultural factors, first and foremostly by language. In the case of the *leges* we are dealing with its textual representations. A man's body exists also, however, in a material and 'objective' way.[149] No historian of culture should underestimate this aspect of the research problem.

An important question for our investigations is the relation between the role which is fulfilled within the researched texts by the 'subjective' concept of the body and the meaning of its 'objective' aspect. We accept the premise that in this relation the culturally formed concept of the body had a overriding and

---

149 Cf. for example, R. Fleming, *Bones for historians: Putting the body back into history*, [in:] *Writing Medieval Biography 750–1250. Essays in Honour of Professor Frank Barlow*, ed. D. Bates, J. Crick, S. Hamilton, Woodbridge 2006, pp. 27–48; the author presents the possibility of recreating on the basis of analysis of burial sites (skeletons and the decoration of graves) the 'bodily' biography of individual persons (their state of health, living conditions, wealth, position within local society, etc.)

'steering' function, i.e. it defined the way by which its material form was presented in the text. Textual manifestations of the 'objective' aspect of the body, which numerically dominate within the barbarian *leges*, were a differential effect of its cultural conception. The 'objectively' existing body was not reflected in the text in a direct way, rather it was formulated by means of the said 'subjective' concept.

The Concept of the Human Body within the Barbarian *Leges*

The concept of the human body was expressed within *leges barbarorum* by the word *corpus* as well as by various Germanic expressions interjected into the Latin text as glosses – *hreo, hreû, hraiwa* (body, corpse). The word *corpus* appears within the *leges* in several contexts and meanings. The most often 'body' is a dead object – a human corpse. In several codes of Germanic laws there are regulations protecting the body of the deceased, both buried and unburied, from desecration.[150] In *Pactus legis Salicae* this rule appears twice – first in the title XIV, 9, where there is talk of 'a dead man' (*homo mortuus*), and the second time in the title LV, 1 and 4, entitled *De corporibus expoliatis*, where there appears the term *corpus occisi hominis*.[151] While in *Lex Salica* in the title XVIII (On a dead man) there is talk of *corpus sepultum*.[152] The Latin word *corpus* corresponds to the first part of the Frankish term *chreomosino*, which appears as a *mallberg* gloss in the titles: XIV, 9 and LV, 1. M. Gysseling has linked this term with the compound *hraiwa-musiđa-, which he explains as a secretively looted corpse, in

---

150 Cf. H. Nehlsen, *Der Grabfrevel in den germanischen Rechtsaufzeichnungen. Zugleich ein Beitrag zur Diskussion um Todesstrafe und Friedlosigkeit bei den Germanen*, [in:] *Zum Grabfrevel in vor- und frühgeschichtlicher Zeit. Untersuchungen zu Grabraub und „hangbrot" in Mittel- und Nordeuropa*, ed. H. Jankun, H. Nehlsen, H. Roth, Göttingen 1978, pp. 107–168.

151 PLSal, XIV, 9: *Si quis hominem mortuum, antequam in terra mittatur, „in furtum" expoliauerit, cui fuerit adprobatum, mallobergo chreomosido sunt, IVM denarios qui faciunt solidos C culpabilis iudicetur;* LV, 1: *Si quis corpus occisi hominis, antequam in terra mittatur expoliauerit cui fuerit adpriobatum, mallobergo chreomosido hoc est, MMD denarios qui faciunt solidos LXII semis culpabilis iudicetur;* 4: [Et antiqua lege:] *Si quis corpus iam sepultum effodierit et expoliauerit et ei fuerit adprobatum, mallobergo muther hoc est, uuargus sit usque in diem illa(m), quam ille cum parentibus ipsius defuncti conueniat, „ut" et ipsi pro eo rogare debeant, ut ei inter hominess liceat accedere* [...]; LXI, 2. *Si guis uero hominem expoliauerit uiolenter.*

152 Cf. H. Nehlsen, *Der Grabfrevel...*, pp. 137 ff.

a way similar to the case of the term *þewa-hraiwa-musiđa ('to loot the corpse of a slave in secret').[153]

In the law of the Ripuarian Franks in title LIV, 1 (*De corporibus expoliatis*) there is mention of the defiling (robbing) of a dead man (*homo mortuus*), while in the title LXXXV, 1 (*De corpus expoliato*) of the stealing of *corpus mortuum*.[154] In Edictum Rothari, in title 15 (*De grabworfin*) concerning the robbing of a grave and a body (human corpse) there is also employed the word *corpus*.[155] In the Saxon laws, and more exactly in *Captulatio de partibus Saxoniae*, title VII, which referred to the pagan funeral service that involved the burning of the deceased's body 'so that the bones turn into ash', corpse is rendered by the phrase: *corpus defuncti hominis*. While in title XXII of this capitulary it is commanded that *corpora christianorum Saxonorum ad cimiteriae ecclesiae deferentur*.[156] This regulation constitutes an example of the direct influence of Christianity up-

---

153 M. Gysseling, *De germaanse*, pp. 81 and 102; similarly K. A. Eckhardt (*Glossar*, [in:] *PLSal*, p. 281) states that "chreumosido" meant 'the looting of corpses' (*Leichenberaubung*); H. Kern, *Notes*, § 90, col. 474–475, § 86, col. 471–472 and § 245, col. 545; "chreomosino" is the distorted form of the word *hreomurdo* or *chreomurdio*, being the compound of two Old Germanic words - chrêo, hrêu (corpse, body) and morđ, (murder) or more broadly a "shameful crime". Chreomurdio was according to him expressed in Latin by the expression *mortui spoliatio*; as well as W. van Helten, *Zu den malbergischen*, pp. 331–332 (§ 63), pp. 341–342 (§ 67) and pp. 350-351 (§ 71) where he writes that the word "chreodiba" (entered into certain manuscripts as "creobeba" or "chreobiba") meant 'the burning of a corpse' (Leichenverbrennung); Cf. *Cap. leg. sal. add. I*, LXX, 1, *Si quis hominem ingenuum* [...] *occiderit et eum ad celandum conburserit* [...].

154 *LRib*, (Cod. B) LIV. *De corporibus expoliatis*. 1. *Si quis autem hominem mortuum, antequam humitur expoliaverit* [...]; LXXXV, 1. *Si quis corpus mortuum, priusquam sepeliatur, expoliaverit* [...].

155 *ERot*, 15. *Si quis sepulturam hominis mortui ruperit et corpus expoliaverit aut foris iactaverit* [...]; In *Edictum Rothari*, title 16, in the heading, there appears the concept of *rairaub*, which contains a word corresponding to the Gothic *hraiwa* (= hrêo) – 'corpse'; cf. H. Nehlsen, *Der Grabfrevel...*, p. 124; cf. also, F. van der Rhee, *Die germanischen Wörter in den langobardischen Gesetzen*, Diss. phil., Utrecht–Rotterdam 1970, pp. 39 ff. and 111 ff.; H. Munske, *Der germanische Rechtworschatz in Berich der Misstaten, philolog. u. Sprachgeogr. Untersuchungen I, Die terminologie der ältrerenwestgermanischen Rechtsquellen*, 1973, p. 266.

156 *Capitulatio de partibus Saxoniae*, title VII; on the subject of Saxon law: G. Theurkauf, *Lex, Speculum, Compendium iuris. Rechtsaufzeichnung und Rechtsbewußtsein Norddeutschland vom 8. bis 16. Jahrhundert*, Köln–Graz 1968, particularly pp. 38–54; G. Ausenda, op. cit., pp. 113–131.

on the content of the law, however equally in this case the word *corpus* was used in the meaning of 'corpse'.[157]

In *Lex Baiwarorum*, in title XIX, 2, 5 and 6 on the desecration of a person's dead body the word corpus was consistently replaced by the term *cadaver*, referring exclusively to a corpse. There is reference here to the hiding of the body of a dead man by throwing it into a river (pt. 2), on the wounding of a corpse by a shot aimed at the birds tearing at it (pt. 5) as well as the conscious mutilating of the body through cutting off its head, feet, ears, or through a blow which resulted in the drawing of blood (pt. 6).[158] Here it is worth emphasising that the word *corpus* was used in Latin, besides its basic meaning of body (of a living person), to refer to human corpses.

In all of the above cases the concept of the body was identified with that of a corpse. Consequently the matter concerned a dead body. Clear is the semantic link between the word *corpus* and the term *homo mortuus*, most pointedly expressed in the formula *corpus defuncti hominis*, or even the identification of these concepts. It can be also seen when we compare the content of the heading *De corporibus expoliatis* and the content of the same norm where one has the term *homo mortuus*. In Bavarian law this similarity in identification occurs through the replacement of the concept of body by the word *cadaver*. We are, therefore, dealing with a characteristic reduction in the meaning of this concept to a single aspect. In this case body is corpse or vice versa, the corpse is the 'proper' form of the body.

A specific situation within which the word *corpus* appears has been described in the regulation of *Lex Alamannorum* (title LXXXVIII) on the causing of miscarriage as a result of beating pregnant women.[159] In the first paragraph

---

157 Cf. G. Theuerkauf, *op. cit.*, p. 41.
158 *LBai*, XIX, 2. *Si in flumine proiecerit. Si quis liberum occiderit furtivo modo et in flumine eiecerit vel in talem locum eiecerit, ut cadaver reddere non quiverit, quod Baiuuarii murdita dicunt, inprimis cum XL solḋ conponat eo quod funus ad dignas obsequias reddere non valet; postea vero cum wergeldo conponat;* 5. *Si vulneraverit mortuum. Et si, ut sepe contigit, aquile vel ceteri aves cadaver repperint et super ad lacerandum consederint, et aliquis sagittam eiecerit et cadaver vulnerverit et repertum fuerit, cum XII solidos conponat;* 6. *Si cadaver leserit. Simili modo quicumque cadaver lederit quem alter interfecit, si caput amputaverit, si manum preciderit, si pes, si aurem, si tantum quod profusionem sanguinis quod reputamus de mortuo, tam minima plaga quam maxima, semper cum XII solḋ. conponat;* cf. A. Niederhellmann, *Artz...*, pp. 133 ff.
159 *LAl*, LXXXVIII, 1: *Si quis aliquis mulierem prignantem aborsum fecerit, ita ut iam cognoscere possis, utrum vir an femina fuisset; si vir debuit esse, cum 12 solidos conponat, si autem femina, cum 24*; cf. cod. B, XCI, 1. *Si quis mulieri pregnanti abortivum fecerit, ita ut iam cognoscere possis, utrum vir an femina fuit* [...]; Cf. also, *LVisig*, VI, 3, 2: (*Antiqua. Si ingenuus ingenuam abortare fecerit.) Si quis mulierem gravidam percusserit*

there is differentiated the amount to be paid in relation to the sex of the foetus. The most important for us is paragraph 2 of this title in which there is defined the compensation for causing miscarriage in a case where the sex of the child cannot be determined. The regulation in question reads as follows: *Si nec utrum cognoscere potest, et iam non fuit formatus in liniamenta corporis, 12 solidos conponat.* (If one cannot determine which of the two [i.e. whether it was a boy or a girl] or still (?) [the foetus] was not formed in its traits (contours) of body, 12 *solidi* are to be paid.).[160] We are here dealing, therefore, also with corpses, but of a specific kind. For the matter concerns an unformed foetus (a human body), not in possession of bodily features although, as one can see, still treated by the law as a human entity.[161]

The concept of the body is here clearly connected with the designation *liniamentum* (*lineamentum*), which meant: feature, trait, contour, shape.[162] The formula *formatus in liniamenta corporis* – even though it was used in relation to an unformed foetus – in fact referred to a formed foetus (*formatus*). The state of formation was described as a contrast to the state of non-formation, and with it initially was the claim *in lineamenta corporis formatus*. From the moment of the creation of features of a body it had crossed over a certain barrier, becoming a shaped entity, whose sex was possible to discern. The body of an unborn child was to gain with this a defined structure. This was to be a defining situation. To-

---

quocumque hictu aut per aliquam occasionem mulierem ingenuam abortare fecerit, et exinde mortua fuerit, pro homicidio puniatur. Si autem tantumodo partus excutiatur, et mulier in nullo debilitata fuerit, et ingenuus ingenue hoc intulisse cognoscitur, si formatum infantem extincxit, CL solidos reddat; si verom informem C solidos pro factus restituat. Cf. M. Elsakkers, *Genre Hopping*, pp. 79–84; eadem, *Abortion*, pp. 237 ff.

160 Deciphering this sentence is not easy, the publisher gives within the critical frame other variants of some of the words: instead of *utrum* in some manuscripts there appears *neutrum* (none of the two), it is, however, not to be excluded that the matter here concerns an incorrect form of the word *uterus*, which means 'foetus'; the least clear in this context are the words *iam non*, in two manuscripts (A8 and B 30) there were replaced by the expression *nec dum* (still not) and *adhuc* (thus hitherto, still/yet).

161 A. Niederhellmann, *Arzt...*, pp. 133 ff. And the footnotes 54–57 and 61. The author draws attention to the fact that the appearance here of the concepts of an unformed and formed foetus appears in ancient texts, first and foremost in Aristotles' philosophy of nature (*Historia animalium* 7, 3). This differentiation appears also in Hippocrates and Pliny the Elder as well as in St. Augustus. She also advances the conjecture that the quoted title LXXXVIII may have been a later addition, something borne out by a comparison with the titles XII *Pactus* and LXX *Lex Alamannorum*, which speak of abortion, in which the two concepts mentioned do not appear as neither does the term 'abortion' (*avorsum*).

162 Du Cange, *Glosarium mediae et infamae latinitatis*, vol. V, Graz 1954, p. 115.

gether with the appearance of the body's contours and appearance of the foetus's sex it entered into a new phase of existence, it gained a new legal status. The body and its sexual features became the criterion upon which the legal situation of the victim was defined. We have mentioned that the above title concerned the bringing about of a miscarriage, i.e. murder. It is not completely clear whether the legislator, in talking about the body of an unborn child, had in mind a corpse or a live foetus. One may, however, assume that the creators of Alamann law related the concept of the body equally to a live foetus. This is, and it follows to emphasis this, a presumption taken from an analysis of a fragment of *Lex Alamannorum*, and not a conclusion drawn *expressis verbis*. More important in considerations over the concept of the body is, however, the differentiation between the foetus in the mother's womb and a born child than the distinction between a live foetus or corpse. This is pointed out also by the fact that in the case of murdering a born child, in a similar way to other persons, the concept of body does not appear in the barbarian *leges*.[163]

If we recognise the correctness of the conclusion that the Alamanns combined the concept of body with the formed human foetus not only with a dead one and compare that fact to the ascertainment on the identity of a human body with the corpse of a dead man, a subsequent thesis may be drawn. We would have then two extreme threshold situations. One, before the child's arrival on this earth, but after the formation of the foetus and the determining of its sex. While the other is when a man was dead. For the corpse was a body which had crossed over a certain threshold, here being the threshold of death. These two situations of the human body, besides the fact that they constitute extremes, have also a common feature: that in both cases the individuals (and precisely its body) found itself as if beyond the circle of social life, remaining in a state of passivity. In the case of buried corpses there exists one more analogy with a foetus – the body in both cases is found in hiding – in the mother's womb or in the earth. The human body as a concept was linked to or identified with threshold situations, with the extreme states of human existence (the beginning and the end),

---

163 Cf. *PLSal*, XXIV, 6: *Si quis uero infantem in uentre matris suae occiderit aut ante quod nomen habeat infra nouem noctibus cui fuerit adprobatum, mallobergo anouuado, sunt, IVM denarios qui faciunt solidos C culpabilis iudicetur*; similarly in the title XLI, 20; *PLAl*, XII: *Si quis mulier gravata fuerit, et per factum alterum infans natus mortuus fuerit, aut si vivus natus fuerit et novem noctes non vivit* [...]; *LAl*, LXX: *Si qua mulier gravida fuerit, et per factum alterum infans mortuus fuerit et novem noctes non vivit* [...]; *LBai*, VIII, 19: *Si quis mulieri ictu quolibet avorsum fecerit,* [...] *Si autem tantum partus extinguitur, si adhuc partus vivus non fuit, XX solidos conponat. Si autem iam vivens fuit, wergeldum persolvat L et III solidos et tremisse*; *LFris*, V, 1 (*De hominibus qui sine compositione occidi possunt*): [...] *et infans ab utero sublatus et enecatus a matre.*

with passivity, the static, with exclusion from social life and the situation of being hidden.

An interpretation of the discussed law *Lex Alamannorum* should take into consideration the fact that it constituted a supplement to another presumably earlier regulation on the causing of death to unborn children (title LXX). It seems rather unlikely that this should have been included within *Pactus legis Alamannorum*, where we can find a regulation very close in form to the recalled *Lex*, i.e. title XII. This results consequently from the fact that the title LXXXVIII had not rather been a part of the former law of the Alamanns which had been registered for the first time in the first quarter of the 7th century. More likely does it seem that it constituted a kind of innovation, one introduced at the time of the creation of the text for the *Lex* (i.e., in the 1st quarter of the 8th century), involving the use of the word *corpus* in a meaning that neither the editors of the code nor legal experts had earlier used. If we were to adopt this hypothesis we would be dealing with an example of the evolution in the categorization of the human body.

In Latin language texts we also have examples of the use of the word *corpus* in the meaning of 'trunk/torso'. In Saxon law (*Lex Saxonum*), in title V one reads: *Si os fregerit vel wlitwam fecerit, corpus vel coxam vel brachium perforaverit* [...]. Similarly in the law of the Thuringia where title X reads: *Corpus transpunctata* [...]. This regulation needs to be examined within the context of those titles preceding and succeeding, in which there is talk of damage to individual parts of the body (e.g., titles 8–9 speak of breaking the bones of a wealthy and free man; in title 11 the stabbing of a leg or an arm).[164] The word *corpus* did not here represent the body as a whole, but actually the 'torso'.

In solving the problem of the means by which the concept of the body is used in the barbarian *leges* it follows to draw attention to the German words that appear in certain Latin language texts of law. For they have a significant meaning for the question here under consideration, of the relation between this concept and the designations for individual parts of the body.

In the oldest Anglo-Saxon code of King Æthelberht (601–604), written in Old English, there is no word designating body. Admittedly in article 61 there appears the term *hrifwund*.[165] The word *hrif* meant, however, not body (like

---

[164] A. Niederhellman (*Arzt*..., pp. 249–250) combines these regulations with the titles of Alamann and Bavarian law mentioned above and explains their meaning as 'perforation of the body'. The Germanic word *wlitiwa* meant a wound resulting in deformation of facial appearance (ibidem, p. 290); cf. Add. Sap. III, 16: *Si ex percussione deformitas faciei illata fuerit* [...] *quod wlitiwam dicunt* [...].

[165] *Æthelberht*, 61. *Gif hrifwund weorðep*, XII *scill'gebete*; Liebermann translated the sentence in the following way: 'If [someone] is wounded in the stomach, 12 shillings in compensation is to be paid'; cf. also F. L. Attenborough's translation (The Laws of the

hreu), but 'stomach'. The context in which it appears (and the very compound) shows that it could not here have represented the body as such, but the torso in the broadest understanding. The most likely meaning of this compound is a wound to the stomach cavity. We come across similar terms in the laws of the Alamanns and Bavarians – *hrevovunt* and *hrevavunt*.[166] The word *hrevo*, which was a component of both of these compounds, constituted a variant of the word *rev(o)*.[167] This in turn meant the stomach/abdominal cavity, then inners of the torso, and consequently a designated fragment of the body, precisely that of the trunk. In both cases they had a very similar meaning to that in the Æthelberht codification i.e. 'wounding of the entrails'.

The absence of terms designating a man's body in the oldest Anglo-Saxon codification is fairly surprising, for in Old English there existed several words designating body with various hues of meaning. The two main ones were: *bodig*, which meant also the trunk and *lîc*. Both were used equally in relation to the body of a living man as to a corpse. There also existed the compound word *lîc-hama* (literally: 'the shell of the body'[168]), referring only to the body of a live person. Besides this there appeared in Old English the words *flæsc*[169] and *hama*, as well as the compound *flæschama*. In addition are known also terms designat-

---

Earliest English Kings, Cambridge 1922, p. 13, who used the word 'belly'. Cf. also A. Niederhellmann, *Arzt...*, pp. 171–172, the author, drawing on philological research, claims that the Old English word *hrif* that meant 'belly, the mother's womb', is derived from an Indo-Germanic root *krep-, *krp- 'body, figure'.

166 *LAl*, LVII, 54. *Si autem in latus punctus fuerit, ita ut interiora membra non contigat.* 55. *Si autem interiora membra vulneratus fuerit, quod „hrevovunt" dicunt* [...]; *LBai.*, IV, 6. *Si cervella in capite appareant vel interiora membra plagata fuerint, quod hrevavunt dicunt* [...].

167 On the subject of the words *hrevovunt/hrevavunt/hreuauuinti* cf. A. Niederhellmann, *Arzt...*, pp. 250–251, he connects *rev(o)* with the Old-Upper-German *href, ref* ('uterus'), the Old English hrif ('stomach') and the indo-Germanic *krep-, *krp- ('body, figure'). The word '*href*', '*hrif*' ('stomach') appears also in the regulations of Frisian law – in the compound *midhrif/midhref*, similarly to the Old English *midhrif*, which most probably meant 'inners/entrails'.

168 *Beowulf and fight at Finneburg*, ed. Fr. Klaeber, Boston 1950, line 1007; An Anglo-Saxon Dictionary, based on the manuscript collections, J. Bosworth, Oxford 1899, p. 636 '*lîc*' whereas this is 'a live or dead body'. Cf. also *Beowulf. Eine Auswahl mit Einführung, teilweiter Übersetzung, Amerhungen und Etymologischem Wörterbuch*, ed. M. Lehnert, Berlin 1967, see the etymological glossary, *lîc-hama*, *lîc* (Leib, Körper), *hama* (Gewand, Hülle).

169 This word appears in the *Laws of Alfred*, XXI, 35 (in: *Die Gesetze der Angelsachsen*, ed. F. Liebermann, Aalen 1960, vol. I, pp. 34–35) in reference to the dead body of an oxen, this code comes, however, from the end of the ninth century.

ing corpses – *hold* and *hræw* as well as the word *limgesihp*.[170] This observation shows that the absence of terms designating the body of a living person was not in the majority of Latin texts their exclusive trait. Taking into consideration the fact that Anglo-Saxon law written down in the first language of the legislators presumably expressed their way of thinking more directly, one may assume that the Latin texts more or less faithfully reflected the thought structures appropriate to Germanic societies expressed through the help of their own linguistic stock.

Some of the crimes against a person's health and life related, potentially, to the whole body. In Latin texts the word *corpus* did not appear in the regulations talking about such crimes. In *Pactus legis Salicae* we have though an interesting *mallberg* gloss defining this type of act. This refers to paragraphs 1, 2 and 3 of title XVII – *De uulneribus*, relating to wounding a man and attempted murder, an attempt at hitting with a poisoned arrow and bloody wounding.[171] In each of these regulations there is the same Frankish designation of the said crimes – *seolandouefa*. The meaning of the term is not clear. Philologists involved in the mallberg glosses have presented various interpretations for it. H. Kern and M. Gysseling have expressed the view that the first component of the compound, *sêla* (**saiwal^o*) meant 'soul', 'life', 'living soul', 'living entity', (cf. Old Saxon *sêola, siole*, Middle German *ziele, ziel*).[172] W. van Helten relates sceptically to reading the said 'seol' in the meaning: 'soul', 'life'. Kern came to the conclusion that the designations *seolan doefa/defa* meant as much as 'to wound with a blow a living entity (soul)'. While Van Helten interpreted the term as 'resulting from a malicious attitude, causing, threatening life'. K. A. Eckhardt in the edition *Pactus legis Salicae* has assumed that the compound *seolando ueua* (*uena*) meant 'a threat to life'.[173]

---

170 F. Thöne, *Die namen dermenschlischen körperteile bei den Angelsachsen*, Kiel 1912, p. 12.
171 *PLSal*, XVII, 1. *Si quis alterum uulneruerit aut uoluerit occidere et colpus praetrfallierit et ei fuerit adprobatum, mallobergo seolandouefa hoc est, MMD denarios qui faciunt solidos LXII semis culpabilis iudicitur, 2. Si quis alterum sagitta toxicata percutere uoluerit et colpus praetersculpauerit et ei fuerit adprobatum, mallobergo seolandouefa sunt, MMD denarios qui faciunt solidos LXII semis culpabilis iudicetur, 3. Si quis hominem plagauerit ita, ut sanguis ad terram cadat, et ei fuerit adprobatum, mallobergo se(o)lando(u)efa, DC denarios qui faciunt solidos XV culpabilis iudicetur.*
172 H. Kern, *Notes*..., col. 477–478; M. Gysseling, *De germanse*..., p. 94.
173 H. Kern, *ibidem*; W. van Helten, *op. cit.*, p. 355, in his opinion *ando* meant 'anger/wrath', 'an angry mood', while *wefa* was to mean 'provoke' or 'punishment for provoking'; K. A. Eckhardt, *Glossar*, [in:] *PLSal*, p. 287; the author assumed that the words *seolando* (*se eo lando*) *ueua* (*uena*) meant 'a threat to life'; M. Gysseling, *De germanse*..., p. 94, gives the following explanations: the compound **saiwala-anđi-webô* may be interpreted as 'attempted murder'[moordpoging]; the part 'ando' (*anđi*) meant

It seems that the more convincing is the explanation of this formula which encompasses the concept of the soul as an element of life. One may cite the regulations of the *Edictum Rothari* to support this thesis, in which there appear two formulas informing about the consequences that could face the perpetrators of certain acts (the planned murder of the king, insubordination against a military commander), and to be precise – *animae suae incurrat periculum* and *sanguinis sui incurrat periculum*.[174] The concept of *anima* ('soul') meant within this context 'life'. This can be especially well seen in title 1 where talk is of plotting *contra animam regis*. The matter here is though that the word *vita* is not used. One can see from this that the Lombards, and presumably also the Franks, willingly used the concept of soul in the context of a threat to life.

In summing up the considerations on title XVII of the Pact of the Salic Franks we note that in the regulations on bodily violation there appeared instead of this a word constituting 'soul' and 'living entity'. The use of this term is puzzling particularly in the case of wounding with bloodletting. For points 1 and 2 concerned attempts at murder. While paragraph 3 referred to a typical crime against the body, wherein there is no talk of any concrete part of the said. This means that wounding could have occurred in almost any fragment of the body. It therefore seems natural that in the Latin or Frankish description of the crime there should appear the concept of interest to us. Yet the Salic codifiers recognised it as a crime against 'the soul' though also against 'a living entity'. Therefore it appears that they had no need for the concept of the body of a living person to edit the regulation under analysis, or more broadly to talk about acts of this kind. The object of crime was man or rather the energy invigorating him

---

in his opinion 'against', while *ueua* (*webô) – 'intention'; *selando efa* (XVII, 1) is 'unsuccessful attempt at murder', *selando ueua* (XVII, 2) – 'attempted murder with the use of a poisoned shot', *selando efa* (XVII, 3) – 'inflict a bloody wound'. Cf. also *PLSal*, XVI, 1 – on the burning of a sleeping man as a result of setting fire to his home.

174 Cf. *ERot*, 1. *Si quis hominum contra animam regis cogitaverit aut consiliaverit*, **animae** *suae incurrat periculum*; 4. *Si quis inimicus provincial intraverit aut introduxerit*, **animae** *suae incurrat periculum, et res eius infiscenntur*; 6. *Si quis foris in exercitum seditionem levaverit contra ducem suum, aut contra eum qui oridinatus est a rege ad exercitum gubernandi aut aliquam partem exercitum seduxerit*, **sanguinis** *sui incurrat periculum*; 7. *Si quis contra inimicus pugnando collegam suum dimiserit, aut astalin fecerit, id est si eum deceperit et cum eum non laboraverit*, **animae** *suae incurrat periculum*; 9. *Si quis qualemcumque hominem ad regem incusaverit quod* **animae** *perteneat periculum, liceat ei, qui accusatus fuerit, cum sacramentum satisfacere et se eduniare* [...] *Et si provatum fuerit, aut det animam, aut qualiter regi placuerit, conponat* [...]; cf. also, title 36. (*animae suae periculum incurrat*); 213. *De crimen adulterii* (*animae suae incurrat periculum*); cf. also title 3.: *Si quis foris provinacia fugire temptaverit, morti incurrat periculum* [...].

which was inseparably connected with the bodily substance. Man was the 'materialised' *seolan* and not therefore an abstractly understood 'body'.

Excursus on the Use of the Word *Membrum*

In discussing the problem of the understanding of the body within the legal regulations cited one needs to refer to the appearance within a part of these of the word *membrum*, i.e. 'part', or rather 'member/component'. In certain cases like in *Lex Ribuaria*, this was used in relation to an unnamed fragment of the body – in title III we read the following sentence: *Si quis alterum in quolibet membro osso* [or: *os*] *fregerit* [...]; in title V, 6 (*De debilitatibus*), there appears a similar use of this term: *Sic in omni mancatione, si membrus mancus* [or: *membrum mancum*] *pependrit* [...]. We must recall, however, that in the Ripuarian Franks legal code there did not appear the concept of a living body. The word *membrum* did not therefore refer to the body in general but to its particular members, that is arms, hands, feet etc.. In the barbarian leges there was not used the formula *membrum corporis*. We see (observe) a very similar practice (situation) in the Burgundian *Liber Constitutionum*, where in title XLVIII (De inflictis vulneribus) we read: [...] *Si quis igitur ictu fustis aut lapidis brachium alterius fregerit et usum eiusdem membri percussus sine cetera debilitatione receperit* [...].

In the law of the Alamanns there was, in turn, used the designation *membra interiora* as the name for internal organs.[175] Also within this context the word *membra* did not refer to the body as such but rather constituted an element of the name that covered the innards understood as a separate 'unit'. The word was not used in this code in relation to other parts of the body. An almost identical formulation appears in *Lex Baiwariorum*.[176] It follows also to emphasise that in both of these cases there appear the same gloss in Germanic designating the name of the damage to the innards – *hrevovunt* (Bavarian law) and *hrevavunt* (the law of the Alamanns). It seems therefore that the explanation of the term *hrevavunt* as 'wounding of the innards' (*interiorum membrorum percussio*), which was given by E. von Schwind, in the explanations to the edition of *Lex Baiwariorum* is justified.[177]

---

175 *LAl.*, LVII, 54. *Si autem in latus punctus fuerit, ita ut interiora membra non contigat, 55. Si autem interiora membra vulneratus fuerit, quod 'hrevovunt' dicunt* [...].

176 *LBai.*, IV, 6. *Si cervella in capite appareant vel interiora membra plagata fuerint, quod hrevavunt dicunt* [...].

177 *Glossarium verborum vernaculorum*, [in:] *Lex Baiwariorum*, p. 487; cf. also: D. von Kralik, *Die deutschen Bestandteile der Lex Baivariorum*, Neues Archiv der Gesellschaft für ältere deutschen Geschichtskunde, vol. 38 (1913), pp. 445 and 447; Th. von Grienberger, Recension zum D. von Kralik, *Die deutsche Bestandteile der Lex Baivariorum*,

It should be emphasised that the designation *membra interiora* had a different character than was the case with the word *hreu*. Though they did indeed refer to the same bodily 'object' they distinctly differed from each other in relation to the means of its categorization. In the case of the said 'internal members' we are dealing with a generalisation, with a depiction devoid of relation to any anatomical fact. While the word *hrêu* or *hraiwa* defines an individual fragment of the body. We are here dealing with two means of perceiving things as well as the formulation of knowledge. The compounds *hrevavunt/ hrevovunt* were presumably terms used by judges or authorities on the laws of the Alamanns and Bavarians. They belonged to the spoken language and represented the typical mentality of an oral culture. *Membra interiora* is rather an interpretation of a German word, which adapted its meaning to the requirements of a thinking developed by the Latin culture of the written word.

There exists within Bavarian law one more example of the use of the word *membrum*. Here the matter concerns the regulation on the killing or striking of a courtyard dog. There appears here also the claim: *Si autem canis per vestimentum aut per membrum hominem tenuerit* [...].[178] This sentence may be translated as follows: 'If the dog were to grab (seize) a man by his attire [clothes?] or by [some] member'. The word *membrum* was here used in a similar way to its usage in the cited regulations of the Burgundian and Ripuarian law, i.e. without reference to a concrete bodily object. The said closer non-defined 'member' related to the concept of 'man'. The concept of *membrum* also constituted its own category in the classification of the objects of crimes against the person. This term functioned therefore as a designation for a separate object. However, it is not visible that the authors of the quoted regulation associated it with the concept of the body.

* * *

In connection with the above analyses of the word *membrum* there arises the question as to the relation between the individual parts or fragments of the body and the way in which the bodily shell of a man was perceived in the majority of *leges*. The material for the solving of this question is the *Lex Baiwarorum* (XIX, 6) regulation, in which there appears the concept of corpses – and therefore the body of a deceased person. There is talk within it of damage/harm to its particular parts. The attempt at solving the fundamental problem is based on the prem-

---

Mittelungen des Instituts für Österreichische Geschichtsforschung, vol. 35 (1914), p. 160. Editors of *Leges Alamannorum*, K. Lehmann and K. A. Eckhardt have accepted that *hrevovunt* meant 'a moral wound' (vulnus mortale), which appears to be an over interpretation; cf. also A. Niederhellmann, *Arzt...*, pp. 171–172 i 247–251.

178 *LBai*, XX (*De canis et eorum compositione*), 9.

ise that the concept of corpse (*cadaver*) may be treated in this case as the equivalent to the body of a living person. The ascertainment concerning the relations between a corpse and its components we shall treat as a lead in our research into the living body. Let us have a look therefore at the content of the relevant rule in the Bavarian legal code: 'If someone were to desecrate a corpse. Similarly, whoever were to damage the corpse [of a man], who another had killed; if he severed his head, if he chopped off a hand, a foot, an ear, if so extremely [he damages] that [there occurs] an outflowing of blood as we calculate for the deceased, both in the case of limited damage as in that of the greatest he will always pay a composition at a rate of 12 solids'.[179]

The body of a murdered man, as results from this sentence, was diminished by the mentioned members. The relation between the concept 'corpse' and the mentioned parts of the body (head, hand, foot, ear) is not, however, clearly marked. One gains the impression that the concept *cadaver* was not of overriding importance in relation to the said parts, but was viewed as an object of the same type. There is no mention of the enumerated members for they were cut off the corpse (body) about which we can read in *Rothari's Edict* in relation to the hand.[180] We are not dealing here with the relation of the whole – parts but that of greater – lesser. Although there do not exist analogical regulations that talk of a living body it seems possible that the way in which its relations were perceived with regard to the members of the body would have been similar to the treatment of a corpse.

The above interpretation may appear unconvincing for it is based solely on an analysis of the language of the *leges*. We shall make reference therefore to a text of a different sort, in which there is equally talk of a man's body and more closely about its individual parts. Here the matter concerns the description of the appearance of Theodoric I, the king of the Visigoths, written by Sidonius Apollinaris, and contained in a letter directed to Agricola.[181]

Sidonius at the beginning of his letter claimed that he intended to characterise Theodoric with regard to his bodily traits – *formae sue quantitatis*. The author also used the word 'body' – *Si forma quaeratur: corpore exacto, longissimis brevior, procerior eminentiorque mediocribus*. (If the matter concerns its shape: the body is elongated, it is shorter than the tallest, [but] taller and more

---

179 Cf. footnote 158.
180 Cf. footnote 182.
181 *Gai Sollii Apollinaris Sidonii Epistula et Carmina*, ed. Ch. Luetjohann, MGH SS vol. VIII, 1, Berlin 1887, I, 1–3; on the subject of Theodoric's letter, cf., H. Wolfram, *History of the Goths*, transl. by T. J. Dunlap, Berkeley 198, p. 206-207; cf. also: A. Loyen, Sidoine Apollinaire, vols. 1–3, Paris 1960–1970; J. Harries, *Sidonius Apollinaris and the Fall of Rome*, Oxford 1994.

strapping than those average). At the end of a detailed listing of the individual parts of Theodoric's body there comes the sentence in which the word *membra* occurs – 'The legs [are] based on tumescent calves which support the mighty members, and [his] feet [are] small'. (*Crura suris fulta targentibus et, qui magna sustentat membra, pes mediocris.*). Admittedly the formula *membrum corporis*, does not appear here, but from the context it results that the word 'members' referred to the concept of the body as used in the text and to the set of its parts. We have therefore in this account a different way of understanding the concept of body and the perception of its fragments than that given in the *leges* quoted. The said *membra* were not something autonomic, as if equally important with regard to the person himself, these were perceived as the 'manifestations' of his body.

The Specifics of Understanding the Concept of 'Body' in Visigoth and Lombard Law

Against the background that has emerged from the analysed legal regulations of the above codification, the understanding of the concept of the human body within the legal codes of the Visigoths and partly also for the Lombards has presented itself as sufficiently different that it justifies separate coverage. In the Visigoth Liber Iudiciorum (VI, 4, 3) we read: *Quicumque igitur ingenuus ingenuum* [...] *turpibus maculis in faciem vel cetero corpore* [...] *maltiose federe* [...] *presumpserit* [...]. In this excerpt the face (head?) clearly distinguishes itself from the rest of the body. The body (corpus) encompassed not only the torso itself but also the limbs and other members. In the self same paragraph there is also talk of the cutting off of any part of the members whatsoever (*quacumque parte membrorum trucidare presumpserit*). Further on the passage concerns the whole body: [...] *ita ut his, qui male pertulertit aut corporis cuntumeliam sustinuerit* [...] as well as: [...] *si qualibet corporis parte servorum truncaverit vel truncare iusserit* [...]. In the last case the matter refers to the cutting off of parts of the body. This same idea was expressed somewhat differently in section V, 5, 13; men were here forbidden to cut off the members of slave men and women from their bodies without court proceedings and clear crimes committed on their part. Here is mentioned in detail the said parts of the body: hand, nose, lips, tongue, ear or foot, eye. At the end there appears a generalisation: *seu quacumque parte corporis detruncaverit.*

It appears that in the quoted regulations of the Visigoth law we are dealing with a different understanding of the concept of 'the body', and for certain with the use of another category of description. It is clearly dealt with here as a whole, which was divided into component parts. Even when the individual parts are enumerated separately as happens in the last of the cited examples, the legislator treats them as components of a body understood in its entirety. Here therefore the body is a compound concept, one made up of its individual parts. They are not the collection of individual objects enumerated without a designation as to the links between them, but the collection of elements relating to a greater whole creating a system. We should also note that in the quoted examples there is talk about the bodies of living people, about an active body, irrespective of the unit operating in society.

In Visigoth laws, as results from the above analysis, the concept of the body of a living person, in a way that is different than is in the majority of barbarian *leges*, was a clearly defined overriding category in relation to its remaining parts. One needs, however, to add that the application here of the practice of listing crimes against the individual members of the body is the same as in the other Germanic laws. The quoted fragments of *Liber Iudiciorum* constitute their own form of introduction containing the general principles of punishing crimes against the body (i.e. various wounds, cuts, beatings, blows and other violations of personal inviolability). The existence of the said general principles bestowed on the Visigoth laws a separate character in relation to those of other Germanic peoples who did not possess a concept of the body thus understood. It follows to add that the lack of a formula for *pars corporis* is especially surprising in the case of the Burgundian legal code, which displays many common features with that of the Visigoths.

A similar understanding of the concept of body, though one not so explicitly expressed as in the Visigoth laws, is to be found in the Lombard *Rothari's Edict*. The word *corpus* appears in several of its regulations. We shall start our discussion from the regulation of damages to the hand. Title 62 (*De incisione manus*) announces that in the case of a hand of a free man being severed the perpetrator was to pay a half of his weregild. If, however, he caused its paralysis, or did not cut the hand off from the body (*et non perexcusserit a corpore*) the fine was a 1/4 of the weregild.[182] From our view point the most significant here are the words *a corpore*. Most clearly the authors of the *Edict* used the word corpus in the meaning of a whole from which one of its parts had been separated. In the

---

182 *ERot*, 62. *De incisione manus. Si quis alii manum absciderit, mediatatem pretii ipsius, sicut adpretiatus fuerit, ac si eum occidisset, ei conponat; et sic sideraverit et non perexcusserit a corpore, quartam partem praetii ipsius ei conponat*; cf. the Italian translation of C. Azzarra (*Le leggi*..., p. 27): "[...] e se la paralizza,ma non la stacca dal corpo [...]", the expression *staccare dal corpo* means 'detach from the body' and the English (K. Fischer Drew, *The Lombard Laws*..., p. 63): '[...] And if it appears that the hand has been paralyzed although not cut off from the body [...]'.

cited regulation the body meant the 'rest' of the members, everything besides the said severed hand. *Corpus* was here perceived as an object which as a result of a crime underwent sectioning – division into 'components'.

In this excerpt there admittedly does not appear the category of parts of the body known from Visigoth law, which would imply the existence of the body as a whole. One may suppose, however, that this idea was known to the authors of the oldest text of Lombard law. It is difficult to explain why this notation related particularly to the hand. It seems that it could have related to almost any part of the body. In other Germanic *leges* there does not appear this sort of scheme in regulations on the cutting off of a hand or other parts of the body.[183] The context in which the excerpt of interest to us is placed inclines one towards the following interpretation: the parts of the body enumerated in *Rothari's Edict* were perceived in the same way as the hand. The body was not so much a collection of individual members as an entire structure composed of elements creating a system.

This idea finds as if confirmation in two other notations in the codification under consideration. Title 382 of the Edict talks about a crime involving the pushing over and knocking down of a freeman. The perpetrator in such a case was to pay 6 solidi fine, in as far as the injured party did not incur 'any other harm to the body' (*alteram lesionem in corpore*).[184] In this regulation we are dealing with a different situation from the previous one and with a different depiction of the structure of the body. For the described act related to no single

---

183 Cf. *PLSal*, XXIX, 1. *Si quis alterum manum uel pedem debilitauerit* [...]; LSal, XLVIII, 2. *Si uero ipsa* [sc. manus] *excusserit, mallobergo chramere* [...]; 9: *Si uero pedis percussus fuerit, mallobergo chuldachina* [...]; *ERot*, 62: *Si quis alii manum absciderit* [...]; 68: *Si quis alii pedem excusserit* [...]; *PLAl*, X, 13: *Et si manus tota excussa fuerit* [...]; XI, 1: *Si quis alteri pedem truncaverit* [...]; *LAl*, LVII, 66: *Si totum pedem absciderit* [...]; *LBaiw.*, IV, 9. *Si quis libero* [...] *manum vel pedem tulerit* [...]; *LRib*, V, 4: *Si manum excusserit* [...], 8. *Si quis ingenuus ingenuum LFris.*, XXII, 27: *Si manus in ipsa iunctura qua brachio adhaeret abscissa fuerit* [...]; *LSax*, XI. *Similiter de manibus, de pedibus* [i. e. abscisis]; *LThur*, XV. *Manus et pes abscisus* [...]; *LFris*, XXII, 34. *Palma manus abscisa* [...]; *Add. Sap.*, II, 6. [...] *si manus abscissa terram cadens tetigerit* [...]; *LAl*, LVII, 39: *Si autem* [brachius] *a cubito absciderit* [...], 40. *Si autem ab scapula abscisus fuerit* [...].

184 *ERot*, 382. *Si quis hominem liberum inpegerit, ut cadat, conponat solidos sex, sic tamen: si alteram lesionem in corpore ipsius non fecerit. Si autem eum inpegerit et non ceciderit, conponat solidos tres*; cf. C. Azarra's translation (*Le leggi...*, p. 103): '[...] a patto che non gli abbia procurato altro danno al corpo'; K. Fischer Drew (*The Lombard...*, p. 128) translates this fragment more or less literally: 'If anyone strikes a freeman so that he falls, he shall pay six solidi as composition provided that he has not caused other injury.'

part but to the body as a whole. On the basis of the first part of the quoted sentence one could assume that its author was thinking not so much about the body as the man. For talk is indeed on *liber homo*. The second part shows, however, clearly that the said pushing and falling over were categorised as bodily harm. This results from the stipulation that the injured party did not suffer 'any other harm to the body'. The said knocking to the floor was therefore perceived as a crime against the body.

The third account confirming the conclusion as to the viewing of the body as a complex structure composed of many component parts is title 27 of *Rothari's Edict*. Its full wording being: 'If someone were to block if he have done him no bodily harm (aliquam lesionem in carne), and if [this] has occurred then 20 solidi [he will pay] for holding [him] and wounding or damage if [it] has been caused according to the below enumerated in this Edict he will pay compensation'.[185]

In this regulation there is employed the designation *lesio in carnem*. The word *caro* means here body. It is a formula corresponding strictly to the designation *lesio in corpore* – from title 382. The difference between the words *corpus* and *caro* were, so it appears, a distinctness in semantic hue. *Caro* accentuated more the material though also the anatomical aspect of the body. For one of the meanings of this word was 'meat'. The said 'damages to the body' are, according to the notation cited, crimes against the body, and more precisely – damages to its various parts listed in the Edict in titles 43 to 73.

In our discussions it follows to also take into consideration the case of the Burgundian *Liber Constitutionum*. According to the fairly uniform view of historians of law who have involved themselves in this codification it displays many features that point to the influences of general Roman Law (particularly *Codex Theodosianus*). On the other hand, however, the concept of the body of a living person and the concept of corpse (corpus, cadaver) do not appear. This is a fairly surprising fact particularly that mention had already been made for there had appeared there the word *membrum* w in relation to injuries resulting from a

---

185 *ERot*, 27: *Si quis homini libero viam antesteterit, viginti solidos ei conponat, sic tamen, ut aliquam lesionem in carnem ipsius non faciat* [...]; cf. Azzarra's translation (*ibidem*, pp. 18–19): 'Se qualcuno per la strada si oppone as un uomo libero, gli paghi una conposizione di 20 solidi, a patto che non gli abbia provocato quale danno fisico [...]'; K. Fischer Drew (*The Lombard*..., p. 57): 'Anyone who blocks the road to a freeman shall pay him twenty solidi as composition, provided no physical injury was caused'; both researchers have used here the category 'bodily/physical harm' (danno fisico, physical injury), which conveys the meaning of the legal term *laesio in carnem*, but change the meaning of the word *caro* from the literal and subject-at-law to a more general and abstract sense.

beating. This question is difficult to explain. The circumstances that could be helpful in addressing the matter is the fact that within the text of Burgundian law we are dealing with an exceptionally unsystematic (or rather chaotic) body of regulations, including those that refer to crimes against the body. Therefore, given that regulations within this area of acts were not created in a holistic way, may be the editors of Burgundian law did not have the opportunity (the need) to use within them the concept of the body of a living man. Such a method did not require the creating of a regulation, for it treated the body as the compound object of crime.

A comparison of the law codes of the Visigoths and the Lombards – on the one hand, and of the remaining barbarian laws – on the other, should be broadened to include one more aspect. Namely, how the matter of protecting human corpses and graves was regulated within *Leges Visigothorum*. In section XI, 2, 1: *De violatoribus sepulcrorum*, there is talk of the robbing of a grave and a deceased person (mortuum), but the word *corpus* does not appear here. Whereas in point 2 *Si sepulcrum mortui auferatur* about the stealing of a sarcophagus[186] there is no mention of the body or of corpses. There also does not exist any regulation which would regulate the matter of body desecration. There is a lack of comparative material which would allow us to ascertain how the Visigoths saw human corpses, particularly within the context of a living body.

The evasion on the part of the Visigoth legislators of resolving the matter strikes one as not being accidental. Given that they had already involved themselves in the robbery and destruction of graves, it would have been natural to create a regulation that protected corpses. Its absence in this situation is fairly surprising. It is difficult to determine what brought about this legislative decision, one thing is at least certain – these laws contain a clearly expressed concept of the body of a living man, yet they are silent on the matter of a corpse. In *Rothari's Edict* there is, however, next to the meaning of the body as a compound whole, the appearance of it, as mentioned, in the meaning of a corpse.

---

186 *LVis*, XI, 2, 1 (*De violatoribus sepulcrorum*): *Si quis sepulcri violator extiterit aut mortuum expoliaverit et ei aut ornamenta vel vestimenta abstulerit* [...]; 2 (*Si sepulcrum mortui auferatur*): *Si quis mortui sarcofacum abstulerit,* [*dum sibi vult habere remedium*] [...]. Cf. H. Nehlsen, *Der Grabfrevel*..., p. 120 ff.; the author interprets the first regulation as desecration of the grave and corpse; in the case of the second point he considers the meaning of the word *remedium*, which could have meant 'medicine', and which could have been linked with the existence amongst the Visigoths of the conviction as to the healing or supernatural properties of objects which had contact with corpses; he finally states though that as the object of the crime was not the corpse itself nor the furnishings of the burial or sarcophagus, the matter concerned the appropriation of someone else's grave and the illegal burying there of a subsequent deceased person.

Other barbarian legal codes – contrary to the Visigoth – do not possess the concept of a living body, certain of them in turn identified the body with the corpse.

* * *

This ascertainment strengthens the conviction as to the separateness of Visigoth laws on the conceptualization of the concept of the human body in relation to the other analysed Germanic laws herein examined. The question arises: how to explain the said difference? This question seems all the more important given that the Visigoths and the Lombards originated from, and belonged to, the same cultural and ethnic circle as the other Germanic peoples. Theoretically, therefore, they could have made use of the same concepts as the remaining Germanic peoples. One may take into consideration two hypothetical solutions:

1) either we are here dealing with essentially the same concept of a living body in all the Germanic codes that is plainly stated in the Visigoth code,
2) or – quite the contrary – the way of understanding this notion contained in the remaining codes was in accordance with a typical view point for the 'barbarians'; while that which is to be observed in *Liber Iudiciorum*, constituted a borrowing of the written word from Latin culture.

The first hypothesis seems unlikely. For one may ask the question: why within the Visigoth codes was it possible to use the designation *pars corporis*, while in the remaining researched legal collections it is not utilised? Did this constitute some difficulty connected with translation from Germanic languages into Latin, editorial reasons, or maybe it was caused by simple chance. It appears that both the number of the said codes as equally the differentiations in the time and circumstances under which they came about exclude the possibility of such an all-encompassing action of incident or set of external circumstances. It is possible, therefore, to consider that the main reason for the absence of the concept of a living human body as an entirety within the analysed legal codes is the fact that the peoples who created the said laws did not utilise in their codifications this type of concept. This happened because it did not correspond to their way of perceiving and conceptualising the bodily exterior of man.

This consequently leaves the second possibility – that the concept of the body as present in *Leges Visigothorum* and *Rothari's Edict* was of non-Germanic origin. This hypothesis seems to be probable. For it is known that the laws of the Visigoths and the Lombards display greater Roman influences than is the case in the others (*Codex Theodosianus*), involving the adoption of concepts and notions, terms and legal institutions. The question as to the Roman providence of the notion of the body in *Leges Visigothorum* and *Rothari's Edict* is obviously a hypothesis requiring detailed comparative research.

Let us refer therefore to the codification of Roman law from the period of the late Empire. The first official collection of imperial constitutions was proclaimed in AD 439. The *Codex Theodosianus* contained 300 acts from the times of Constantine the Great.[187] The Theodosian Code played a huge role as a source of knowledge about Roman law in western Europe. The second, much broader and more perfected codification initiated by Emperor Justinian (consisting of: *Corpus iuris civilis*, *Digesta*, *Institutiones* and *Novellae*), was extremely poorly known in the West during the period of the early Middle Ages and consequently did not have an influence on the knowledge of Roman law amongst the creators of the Germanic legal codes.[188] Besides, in the first quarter of the 6th century there came into being two collections of Roman law, compiled upon the initiative of the Germanic rulers – the king of the Burgundians, Gundobad (*Lex Romana Burgundionum*) and the Visigoth Alaric II (*Lex Romana Visigothorum*, called Alaric's Breviary), which preceded chronologically the Justinian codification.[189]

Therefore, the aforementioned legal codes could be the sources from which the Germanic codifiers drew their knowledge as to the Roman understanding of body as a subject of the law, in particular penal law. An examination, however, of the content of the said monuments to Roman legislation allows one to state that their influence in this area is fairly problematic, if not imperceptible. For crimes against the body were formulated without recourse to words which expressed the concept of the body itself.[190] It appears within them in connection with punishments for certain crimes, but in this case its meaning was not ex-

---

187 *Theodosiani Libri XVI cum Constitutionibus Sirmondianis*, ed. T. Mommsen, P. M. Meyer, vol. I, Zurich 1905; Cf. I. Wood, *The Code in Merovingian Gaul*, [in:] *The Theodosian Code*, ed. J. Harries, I. Wood, New York 1993, pp. 161–177; cf. also, H. F. Jolowicz, *Historical Introduction to the Study of Roman Law*, Cambridge 1952.

188 *The Digest of Justinian*, ed. Th. Mommsen, P. Krueger, A.Watson, vol. IV, Philadelphia 1985; *Corpus Iuris Civilis: Codex Justinianus*, ed. P. Krueger, Berlin 1900; I. Wood, Code..., p. 163; R. Collins, *Law*..., pp. 5–7; M. Sczaniecki, *Powszechna historia państwa i prawa*, Warszawa 1997 (9th edition), p. 42.

189 *Breviarium Alaricianum. Römisch Recht im Frankischen Reich*, hg. M. Conrat (Cohn), Leipzig 1903; *Lex Romana Burgundionum*, [in:] *Leges Burgundionum*, MGH LNG, vol. II/1, ed. L. R. von Salis, Hannover 1892; cf. M. Sczaniecki, *op. cit.*, p. 72; I. Wood, *Code*..., p. 162; H. Siems, *Lex Romana Visigothorum*, [in:] *HRG*, vol. II, col. 1939–1949; H. Nehlsen, *Lex Romana Burgundionum*, [in:] *HRG*. vol. II, col. 1927–1934.

190 Cf. *Lex Romana Burgundionum*, V, 1: *Si quis forte ita temerariae admittitur, ut vulnus aut fractura ossum infligatur* [...] *solutio vel vindicta facti ipsius pro qualitate persone in iudiciis arbitrio estimatione consistit, secundum regulam Gaii sub titulo de iniuriam accione* [...].

pounded upon.[191] In late Roman common law the regulations on crimes against health, acts of physical violence were formulated in a completely different way than the regulations known from the *leges* of the Germanic peoples, which dealt with these questions. There was absent within them also regulations on the whole body as equally those dealing with the large number of individual body parts and components.[192] The concept of the body appears in the Justinian *Digests* in the regulations on the desecration of graves and corpses.[193]

These observations lead one to the conclusion that the codifications of Roman law from the 5th and 6th centuries rather do not constitute the source from which the Visigoth and Lombard legislators could have derived their concept of the body as a compound whole. It seems, therefore, that the pattern in question has to be looked for elsewhere. This conjecture directs our attention towards an-

---

191 *Ibidem*, VII, 1: *Si quis ingenuus ingenuo crimen intendens, quod obiecerit, se scripserit probatum, si probatio defuerit, inscribendi se cum eo, quem accusat, corporali supplicio licentia non negatur, ita ut aut caput aut facultatem suam obligat, sicut lex Theodosiani libro nono sub titulo primo designat [...]*.

192 For the only crime against a man's health mentioned in Roman law from the Empire period, which involved the damage of a concrete part of the body, was castration. What is of interest though is that the genitals (and particularly the testicles) were not directly named in the rules regulating this case; cf. *The Digest of Justinian*, vol. IV, p. 820, XLVIII, 8, 4: *Idem divus Hadrianus rescripsit: Constitutum quidem est ne spadones fierent* [...] *et in servos, qui spadones fecerint ultimo supplicio animadversendum esse* [...] *Plane si ipsi, qui hanc iniuriam passi sunt, proclamaverint, audire eos praeses provinciae debet, qui virilitatem amiserunt* [...]; 8, 5: *Paulus libro secundo de officio proconsulis. Hi quoque, qui thlibias faciunt, ex constitutione diui Hadriani ad Ninnium Hastam in eadem causa sunt, qua hi qui castrant*. In relation to male genitalia there appears here the description of limited precision *virilitas* (masculinity, fertility), while the word testicles (*testiculi*) does not appear at all. On the subject of these regulations compare K. Amielańczyk, *Rzymskie prawo karne w reskryptach cesarza Hadriana*, Lublin 2006, pp. 141–148, and particularly footnote 46, in which the author explains the concept of *castratio* (cutting off the testicles with sword), *thlibia* (the squeezing or crushing of the testicles) and *thlasia* (a hemlock compress on the testicles), referring to the work of D. Dalla, *L'incapassita sessuale in diritto romano*, Milano 1978, p. 48; In the codifications of Roman law used in the Germanic kingdoms (*Breviarium Alarici*; *Leges Romana Burgundionum*, [in:] *Leges Burgundionum*, pp. 123-163) there is even an absence of such laconic references to the body as the subject of crime.

193 *The Digest of Justinian*, vol. IV, D 47, 12, 7: *Marcianus libro tertio institutionum. Sepulcri detriorem condicionem fieri prohibitum est: sed corruptum et lapsum monumentum corporibus non contactis licet reficere;* 11: *Paulus libro quinto sententiarum. Rei sepulchrorum uiolatorum, si corpora ipsa extraxerint uel ossa eruerint, humilioris quidem fortunae summo supplicio adficiuntur, honestiores in insulam deportatur. Alias autem relegantur aut in metallum damnantur.*

other type of text that arose during the epoch under examination. One may consider the works of the Christian thinkers with academic, philosophical and moral reflection upon the state of man's body and its component members. An example of this being Cassiodorus's *De anima* and *Etymologiae* by Isidore of Seville.[194] These authors, although expressing a Christian point of view, have drawn widely from the intellectual achievements of Roman culture, and particularly its decadent phase. Both make reference to St. Ambrose's *Hexameoron*, the bishop of Milan of the end of the 4th century.[195] A somewhat different current of Christian thought within which the problem of the body is also examined is monastic literature, and particularly the rules of particular orders, for example: the Master's Regulations or the Regulations of St. Benedict.[196] The intellectual tradition which expressed the body as an autonomous subject while at the same time an object composed of various members, was continued also during the final stage of the Caroline period. An example of this phenomenon is the reflection on the movement and gestures of the body of Remigius of Auxerre. In the work *Commentarium in Martinum Capellam* it contained the following sentence: *Inter motum et gestum hoc distat quod motus est totius corporis, gestus proprie manuum vel ceterorum membrorum.*[197]

The perceiving of a man's body as a compound whole could have had its origin not only in the Roman intellectual tradition but also in evangelical thinking. Here the matter concerns first and foremost St. Paul's reflections on the body of Christ as a metaphor for the Church, these being in his First Letter to the Corinthians: 'For Christ is like a single body with its many limbs and organs, which, many as they are, together make up one body [...] If the whole were one single organ, there would not be a body at all' (I Cor 12. 12, 14. 19).[198]

---

194 Cassiodorus, *De anima*, chp. IX, *De positione corporis*, Patrologia Latina 70, col. 1295–1298; Isidorus Hispalensis, *Etymologiae*, XI,1, *De homine et partibus ejus*, Patrologia Latina, 82, col. 65.
195 Cf. J.-C. Schmitt, *La raison de gestes dans l'Occident médiéval*, Paris 1990, p. 64-65.
196 *La Régle du Maître*, edited and French translation, A. de Vogüé, Paris 1964, 7, 12: '[...] we must with the whole of our body and with all the strengths of its members adapt ourselves to the service of the Lord [...]' quoted after : J.-C. Schmitt, *op. cit.*, p. 79.
197 Remigius Autissiodorensis, *Commentarium in Martinum Capellam*, ed. C. E. Lutz, Leiden, vol. 1, I, 37, 7, p. 136.
198 *Biblia Sacra. Vulgatae editionis*, ed. A. Colunga, L. Turrado, Madrid 1999, pp. 1115–1116: *Sicut enim corpus unum est, et membra habet multa, omnia autem membra corporis cum sint multa, unum tamen corpus sunt: ita et Christus.* [...] *Nam et corpus non est unum membrum, sed multa.* [...] *Quod si essent omnia unum membrum, ubi corpus?* [...].

The Means by Which the Collection of Regulations on Crimes against Parts of the Body Were Composed.

Our research has hitherto concentrated itself around the problem of understanding the concept of body in barbarian *leges* texts, and therefore with what we call the 'subjective' or intersubjective aspect of the problem of interest to us. We have examined the body as a cultural phenomenon. The search for these in no way exhausts the scope of the problem area, which derives from the reading of the regulations on crimes against the body. At present our intention is to examine the content of the said regulations from a different viewpoint. The problem we would like to explore is the presence within the regulations of the *leges* of elements of transfer which speak about an 'objective' body – i.e. about the defining of individual parts of the body or 'bodily objects'.

In analysing the regulations that talk of crimes against the body we find ourselves confronted with a paradoxical situation: for on the one hand it results from the presented source material that the majority of German codifiers connected the concept of the body first and foremost with corpses (and not with a living person) or with the trunk. On the other hand the codes of Germanic laws contain many regulations on violation of particular human parts which (besides the Visigoth and Lombard laws) were nowhere referred to or called parts of the body or body itself. However, here there appear a mass of names for organ /members of the body: its external and internal organs. The body 'disintegrates' in Germanic legislation into dozens of composite pieces, which were not combined into a whole.

In the majority of *leges* the relation between the said parts of the body and the body as a whole were not clearly defined. For sure, however, they were not formulated according to the scheme of entirety - parts, for here there did not appear the actual word 'body'. The fact that the relation in question was not presented by means of defined categories (body – whole – part) does not mean, however, that the said components, which we treat as parts of the body, were presented within the legal works under examination as a collection of an undefined structure. It equally seems that we are here dealing with another means of perceiving the concept of the body, its parts and their mutual relationship. In order to present the arguments backing this hypothesis we shall examine collections regulating incidents of crimes against bodily integrity and inviolability, and particularly their composition, and more exactly the order in which they are mentioned.

In the Laws of Æthelberht the block of points on bodily harm are covered by regulations 33 to 72, where there is talk of damage/harm to 39 parts of the body and also bodily objects.[199] They were mentioned in the following order: hair (33), skin on the head/scalp (34), the skull bones (35), the cerebral leptomeninx (36–37), the shoulder (38), the ear (39–42), the eye (43–44), the mouth (44), the nose (45 and 48–49), the cheek (46) both cheeks (47), the lower jaw (50), teeth: front teeth, middle teeth, back teeth, further teeth (51), the organ of speech (the tongue?) (52), collarbone (52, 1), arm (53; 53, 1), thumb (54), the thumbnail, the index finger, the middle finger, the ring finger, the little finger (54, 1–5),the nail of each finger (55), the face (56), the nose (57–58, 1), an undetermined part of the body (59–60), innards (abdominal cavity) [*hrifwund*] (61; 61, 1 and 63), genitalia (64; 64, 2), the hip bone (65), the hip (67), ribs (66), femoral tendon (68), the foot (69), the big toe (70), every other toe (71), the big toe nail (72), the toe nails of the other toes (72, 1).

As can be seen, the collection of regulations given was compiled according to an easily recognised scheme of things 'from top to bottom'; the various parts of the body are enumerated from the hair to the toes. There exist obviously certain departures from this model: between the cerebral leptomeninx and the ear there unexpectedly appears the shoulder (38), after the finger nails there appears the face and nose (56–58), while between the regulations on the hip we read about the ribs (66). Besides these four notations which were, as it appears, rather the result of an oversight and the supplementing of this list than something resulting from some intended action, the indicated sequence of the appearance of the individual parts or bodily objects has been preserved.

In *Rothari's Edict* crimes against the body are equally enumerated according to the body scheme visible from head to foot. Here 22 parts of the body appear, not counting those titles which referred to damages or violations in the area of the body. Crimes are here described in the items 43 to 73. Item 43 concerns general wounding and bruising caused in a scuffle, while the next the inflicting of a punch to an unnamed part of the body and also slapping the face (*alapa*). The subsequent paragraphs refer to violation of the scalp, skull, eye, nose, lips, front and side teeth, the ear, the whole face, again the nose and ear, the arms, trunk (ribs), the hip, hand, thumb, remaining fingers, foot and toes.

As can be seen the regulations drawn up for bodily damage refer to the scheme of the human figure. Its survey was done by moving fairly consistently from top to bottom. The only small disturbance in this order involves the fact that after describing the damages to the various parts of the face there is again

---

199 The translation and interpretation of the individual regulations on the basis of the edition of Kentish laws by F. L. Attenborough, *The Laws....*, p. 9-15 as well as L. Oliver, *Beginning...* (*Appendix II. Comparison of Retribution According to Amount in Æthelberht*), pp. 195-198

reference to the nose and ear, which earlier had been mentioned in the paragraphs on their cutting off. Next there is separate treatment given to their wounds. It follows to add, however, that this departure from the vertical order of things did not break another principle visible here, that is the division of the body into segments on a horizontal plane. The paragraphs concerning the wounding of the nose and ear appear before those that relate to the arm and trunk. Therefore the group of regulations on the head and face create a separate whole. We can observe the principle of division into segments in one more place, i.e. after the paragraph talking about damage to the hip there occurs a transfer to the regulation dealing with the hand, which in a man standing finds itself at the self same height. Further, consequently, are enumerated the paragraphs on the feet and toes.

In *Lex Alamannorum* the regulations on crimes against the body were located in the title LVII containing sixty nine paragraphs.[200] The first two speak of the striking-wounding of the body without defining the parts of the body they refer to. Potentially these could refer to the whole body. The subsequent four paragraphs (3-7) concern the head, the skull and the brain. Subsequent talk is of damage to the ear (8-10), the upper and lower eyelid (11-12), the eye (13-14), the nose (15-17), the lower and upper lip (18-19), the upper front, lower front and other teeth (20-23), the tongue (26) and wounds to the face (27) there where it is not covered by hair or stubble. Further there are mentioned regulations on the cutting through of the neck (28) so as to subsequently return to some extent to the area of the head – here are located points on the shaving off of hair and the shaving of beards (29-30). Subsequent paragraphs related to various damages to the arm, the elbow and the hand (31-40). Later there is talk of various parts of the fingers (41-52) and the causing of paralysis to four forefingers (53). The next regulations (54-57) deal with various damages to the torso and innards. The next two paragraphs (58-59) are devoted to damage to the genitals.

The Frisian law is, against the background of other codifications, a fairly specific case in relation to the aspect discussed. This is the result of two features. The XXII title of *Lex Frisionum* equipped with the heading *De dolg* (On wounds) contains the most extensive list of damaged parts of the body. This is composed of 89 points, which in the main precisely regulated cases of the mutual wounding and violation of the body or damage to the organism's functioning. Besides, however, the edition of the Frisian code covers a part entitled *Additio Sapientum*, which contains two titles, equally the types of crime of interest to us.

---

200 This arrangement also functioned in Alamann law already under its initial codification i.e. in *Pactus legis Alamannorum* that came into being in the first quarter of the 7th century cf. *PLAl*, I–XII.

Title II of *Additio* covers 10, while III – 78 paragraphs (Add., II–III), and therefore in total almost the same number as title XXII.

The arrangement of the title *De dolg* points also to one more specific trait. The paragraphs 1-65 concern the damage to various parts of the body arranged, with few exceptions, in order from head to foot; points 66-75 talk of the quantity of wound size and the number and magnitude of the fragments of broken bones, 76–89 again concerns individual parts of the body enumerated without a designated order. Here are to be found points (82–83 and 88–89) concerning the violation of inviolability, including also that of a woman's body.

> The head (hearing) (1–2), skin on the entire body (3–4), skin on the head (5), the skull bone (6), the cerebral leptomeninx (7–8), the ear (9), the nose (10), the highest, middle and lowest wrinkle of the forehead (11–13), the eyebrow (14), the eyelid (15), the nose (16), moustache (17), cheek (18), front tooth (19), corner tooth (canine?) (20), molar tooth (21), collarbone (neck?) (22), ribs (23), arm bone above the elbow (24 and 26), arm bone below the elbow (25–26), hand (27), thumb (28), index finger (29), middle finger (30), ring finger (31), little finger (32), five fingers (33), hand without fingers (palm) (34), the highest, middle, lowest joint of one of the four forefingers (35–37), the wrist (38), elbow (39), the shoulder joint (40), the highest thumb joint (41), the lower thumb joint (42), the 3rd thumb joint (43), the elbow, the shoulder (44), the eye (45–46), the chest/breast (47), the pericardium (48–49), the membrane between the liver and the spleen (50–51), the belly (52), the intestine (53), the stomach (54–55), the intestine (56), the penis (57), testicle (58), the scrotum (59), the thigh (60), the calf (61), the whole foot (62), the big toe, each of the remaining toes, a toeless foot (63), the toe joints (64), hair (65), skin on the whole body (66–70 and 75), bones (72–74), the hand (76), the arm (77–78), the lung (80), the back (spine)? (81), grabbing and bounding, casting into deep water (82–83), arm and leg (84), both cheeks and tongue (85), both legs and scrotum (86), a woman's breast (88), the vulva (89).

In a similar way to the above mentioned codes, the order of enumerating crimes according to the scheme of 'top – bottom' was disturbed in several places. This is visible in the case of points 38-40 on the wrist, elbow and shoulder, which appear after those points dealing with the fingers and arm, and between the notations relating to the fingers; similarly in paragraph 44 and 45 there is talk again of the elbow and shoulder, in 46 on the eye. Besides, following on from the toes (63-64) there unexpectedly appears hair (65). A separate case concerns points 76–81 and 84–89, which contain the regulations supplementing the cases missed in the first part of the list.

Titles II and III of *Additio Sapientum* contain regulations that to the greater degree concern damage and violation of those self same parts of the body about which mention was made in title XXII. The basic difference involves the fact that in *Additio* they are enumerated in a different order. Title II (1–10) refers in succession to the hand, fingers, fingerless hands, 3 parts of individual fingers. Title III contains 78 items. The order in which they appear does not indicate the application of a definite single ordering scheme. One may, however, point here to groups of regulations that constitute their own form of block, which concerned one fragment or a region of the body. For example – §§ from 1 to 7 talk about damage to the foot and toes, §§ 8–23 – about the head and parts of the face, 24–27 – on damages to unspecified bones, 28–31 – ribs and innards (stomach, intestines), 36–38 about teeth, 39–40 again about the head, 42–43 and 46 generally about skin, 47–48 – about the eye; 49–58 – about the measure of wound length. Points 59–65 and 74 talk about parts of the head (the eye, nose, cheek, tongue), the testicles and arm.

A comparison of the content of the individual notations of title XXII and the titles of *Additio* mentioned show that despite a changed order in the enumeration of crimes against the successive parts of the body, they display a noticeable similarity with regard to the content of individual regulations. The majority of these describe the same damages to the same parts of the body. There are, however, visible certain differences, e.g., in *Additio* there exist regulations on the cutting off of parts of the toes, something which is absent in the *De dolg* title; in *Lex Frisionum*, XXII there is a rule that talks generally about breaking the nose, while in *Additio* three cases are taken into consideration: breaking the nose on the one side, breaking the nasal septum (*cartilago*) and breaking it right through (crosswise); in the group of regulations on damage to the face and head there also appear those which have no equivalent whatsoever in title XXII – the deforming of the face (so called *wlitiwa*) visible at a distance of 12 feet (approx. 3.5 metres) as well as crushing (contortion?) of an eye or the mouth/lips (ut torqueantur), wounding of the head causing hypersensitivity to warmth and cold.

It is not easy to explain what is the cause of the differences between the list of regulations in title XXII and those in *Additio*, particularly that they differ also in relation to the value of the punishments earmarked for the same types of crime. If one were to adopt the thesis of H. Siems that *Additio* was in essence a working notation of an oral legal transfer, while *Lex* (including title XXII) constituted the subsequent radically processed version of the code's text, then it is not possible to state upon this basis which of the two lists of crimes against the body would most correspond to the ideas and also cognitive models of the authors of the Frisian *lex* on man's body as the object of crime.

The quoted law regulations of the Anglo-Saxons, Lombards, Alamanns and Frisians constitute an example of the most consistently applied scheme of the human figure viewed from head to foot. This also appears in other codes (the Visigoth *leges*, that of the Ripuarian Franks and of the Saxons), although not in such a 'pure' form.

In the case of the Visigoth *Liber Iudiciorum*[201] the collection of regulations of interest to us on crimes against the body, and more specifically damages and violations of various kind, was inserted in the 6th book *De sceleribus et tormentis*, in title 4 *De contumelio, vulnere et debilitatione hominum* (On insults, wounding and disabling people), which in K. Zeumer's edition of *Lex Visigothorum* this was headed: *De vulnere et debilitatione*. Title 4 is composed of 11 points. From the viewpoint of our investigations the most significant are paragraphs: 1. (*De cedibus ingenui adque servi*) and 3. (*De reddendo talione et copositionis summam pro non reddendo talione*), which combine the descriptions of crimes against the body.[202] The individual parts of the body appear here in the following order (pt. 1): the skin on the head, the skull bone, (pt. 3): the hair, face, whole body, head, eye, nostril, nose, lips, ears, hand, thumb, index finger, middle finger, ring and little finger, toes, teeth, ribs.

The regulations on crimes against the body in *Lex Ribuaria* were located within a group of regulations constituting a cohesive sequence (titles I-VI).[203] These titles cover the following parts of the body: the body in general (I-II), a single bone (III), the torso between the ribs (IV), the ear, nose, eye, hand, thumb, index finger, foot, any toe (V, 1–10), the genitals (VI).

In *Lex Saxonum* the regulations on crimes against the body were contained within the titles from I to XII.[204] The order in which the individual crimes are enumerated is fairly detailed. The first four titles and the first point of the fifth talk about damage which could affect any part of the body.[205] Titles V, 2-10, XI and XII concern, however, definite parts of the body, which are enumerated in

---

201 This code comes from AD 654. The majority of the regulations contained within it (around 3/5th) derive presumably from the first writing down of the laws ascribed to Euric (the end of the 5th or beginning of the 6th century), although possibly a part of them were corrected by Leovigild (569-586); cf. R. Collins, *Visigothic Spain...*, p. 235-236; there is also an account of the subject literature; cf. P. D. King, *Law and Society...*, p. 20.

202 These are (pt. 1): head injuries: bruising, cutting the skin, wounding to the bone, bone damage, (pt. 3): hair removal, body and facial wounding, striking with a whip or bat, slapping, bloodless striking of the face, black eyes, cutting the nostrils or nose, disfiguring the nose, lips and mouth, injury to the hand (cutting, bringing about paralysis), cutting off the thumb, index finger, middle finger, ring and little finger. Similarly to the Burgundian codification these regulations referred both to freemen as equally to slaves.

203 These titles refer to freemen, apart from these there exist in this code a second analogical section relating to slaves (titles XX – XXVIII).

204 These refer, differently than in other *leges*, to individuals belonging to the magnate social stratum.

205 *LSax*, I. hitting, II. bruising or causing swelling, III. blood wound, IV. wounding to the bone, V. 1. bone. These resolutions were fairly often applied in barbarian *leges* cf. *ERot*, 42-44; *PLSal*, XVII, 1-2; *LAl*, LVII, 1-2; *LBaiw.*, IV, 1-3.

order from head to toe: (V) bone, face torso, leg, arm, (XI) one eye, both eyes, one or both ears, nose, hand, foot, one testicle, both testicles, (XII) ear, eye, nose, hand, foot. Title XIII referred to the fingers and toes: a whole thumb, half a thumb, the whole of a little finger, $1/3^{rd}$ or $2/3^{rd}$ of a little finger, a whole index finger, $1/3^{rd}$ and $2/3^{rd}$ of an index finder, a whole middle and ring finger. Given this background the content of titles VI-X is fairly surprising, for here talk is of physical violence but, with the exception of cases of grabbing by the hair, they refer to the body only indirectly.[206] In the researched code we are consequently dealing with the combination of two schemes wherein the order by which crimes are enumerated are defined; the first of which was based on a specific classification of the types of damage or violation, while the second refers to the bodily arrangement from head to foot.

* * *

The above analyses of the compositional arrangements in the regulations on crimes against the body constitute material for deliberations on the subject of perceiving the collection of members of the human body through an analysis of the *leges* in their written form. At the start it follows to ask the question as to the mechanism by which the said arrangements arose. For an explanation of this question is necessary to define the means by which the parts of the body were categorised. We shall therefore attempt to reconstruct the means of creating the said collections of regulations. The fact that in the cited codes the damaged parts of the body are mentioned on the basis of an arrangement from head to toe shows that the individuals editing the said regulations took man's body as their structural model for the regulations under examination. In composing and compiling the said collections the editors noting down the law 'reproduced' as if the arrangement of the body members/components. This statement is of a fundamental significance for us. For it defines the means of understanding the creators of these regulations and the categories they utilised. The cognitive starting point for them was not an abstract concept of the body, structuralised according to the scheme of whole – part, but was known to them from their everyday experience of the figure of man.

The principle feature of the collections under consideration was, however, the fact that the cognitive model which was applied by their editors was not directly indicated but rather was to be found beyond the texts. They supposed in this that the recipients (readers or listeners) of the regulations would equally be able to utilise this self same external context. The category to which the said edi-

---

206 *LSax*, VI. cutting through a garment or shield with a sword, VII. grabbing by the hair, VIII. an attack with a sword in hand, IX. throwing from a bridge, boat or shore into a river without drowning, X. immersion by force in water

tors were referring was knowledge as to man's bodily structure. This had, however, a specific character – for it resulted from everyday experience and from practical thinking, and not from a formal reasoning. The argumentation supporting this thesis is the way the content of the regulations is formulated. It follows to here indicate the practice widely used in *leges barbarorum*, which as a result of its characteristics may easily escape the attention of their present-day reader. Here the matter concerns the fundamental fact that the analysed acts on bodily damage mention as their object man. He was the object of criminality, constituting a broader category than a particular part/fragment of the body.

The structure joining the said members was the human form. In this understanding the said 'parts' created 'a whole'. In the understanding of those who wrote down the law these constituted, however, individual 'objects' belonging to man. In this light the status of the body itself as a concept is presented fairly unclearly. It seems that the creators of the said regulations did not require this kind of conceptual category. They perceived the phenomena which were the subject of their interests in a simple way, while one at the same time objective - they saw on the one hand the man, while on the other his nose, finger or foot. For such a mentality the figure of man became a useful regulatory model. As we shall prove later on this was not adopted as a automatic compositional scheme for the regulations discussed.

* * *

The textual scheme reconstructed above, according to which regulations on crimes perpetrated against the body were written down, was not applied in all the barbarian codes. We shall now examine those codifications in which other arrangements existed for the regulations here under consideration. In *Lex Baiwarorum* the above described scheme for the listing of crimes against the body in an order from head to toe was changed to a large degree, so that it is consequently difficult to talk in this case of its application. Admittedly in a few of the first titles there is talk, in a similar way to other codes, of crimes against the whole body as well as about various forms of damage to the skull, knocking out the eye, then, however, there occurs a title on the hand and foot as well as a repeat reference to the eye. The subsequent three titles concern the fingers. While following this there is talk of damage to the arm, before the code turns to damage to the face: the nose, ears, lips, eyelids, teeth. These regulations end with a set of text devoted to damage to the body.[207] A somewhat similar situation is to be found in the case of the Frankish *Pactus legis Salice* and *Lex Salica*.

---

207 Cf. *LBaiw*, IV, 1–16. Here is their list: hitting an unspecified fragment of the body (1-2), the cutting of an unspecified vein (artery), cutting the skin on the head, the breaking of the skull bone, hitting the head (4), breaking the open bones of the skull or the bones

In the codes of Salian law, crimes against the body were listed according to a fairly complex and somewhat unclear order. They were incorporated within two sections – XVII. *De uulneribus* (On woundings) and XXIX. *De debilitatibus* (On disabilities). The very headings for these sections suggest that these crimes were divided according to the criterion of types of bodily damage. Significant is the fact that in the chapter *De uulneribus* there is chiefly talk on various types of wounding, while in the title *De debilitatibus* on cases of parts being severed resulting in paralysis or the cutting off of various parts of the body. It seems, however, that the Frankish codifiers used also other schemes in ordering the listing of crimes against the body. We shall examine in what way the regulations contained in each of the titles is arranged. In the chapter *De uulneribus* the three first paragraphs concern wounds which must have been inflicted on the victim's whole body. For there are not specified any particular parts of the body. Points 4 and 5 talk of serious violations to the skull (resulting in the appearance of the brain or the falling from the wound of three bone fragments). Paragraphs 6 and 7 refer to torso wounds i.e. wounds between the ribs and in the belly with the violation of the intestines. Points 8, 9, and 10 concern blows and wounds that could potentially happen on the entire body (the matter concerns blows with a stick or fist, as well as blood wounding).

The title *De debilitatibus* is of a somewhat different character. Talk is here exclusively about damage to specified parts of the body. The first point of this chapter is especially interesting for us. For it deals with damage to five parts of the body: the severing of the hand (*manus*), the foot (*pes*), ear (*auriculum*), nose (*nasum*) and gouging out an eye (*oculus*). Significant here is the order with which they are listed – hand, foot, eye, ear, nose. The significance of this paragraph makes itself known while reading further points. Paragraphs 2 and 3 concern other damage to the hand, 4 and 5. Talk about the thumb, while points 6 to 9 other fingers. Paragraphs 10 and 11 concern the foot, 12–14 in turn – the eye, the nose and the ear.

It appears, therefore, that the content of point 1 was of a 'scheduling' nature – it designated the order for the appearance of the various parts of the body in

---

of the arm above the elbow (5), exposing the brain or innards (6), illegal bounding with rope (7), other tying up (8), knocking out the eye, cutting the hand or foot (9), the causing of paralysis to the eye, a foot, or hand (10), the cutting off of a thumb, index finger or little finger, cutting off the middle fingers, causing paralysis to the fingers (11), the breaking of the arm above and below the elbow (12), piercing the nose (13), puncturing the ear, cutting off the ear, the causing of deafness, disfiguring the ear (14), the deforming of the lower lip and lower eyelid, wounding the lower eyelid causing watering of the eyes, the wounding of the lower lips resulting in bruising, deforming the upper lip or the upper eyelid (15), the knocking out of a molar, another tooth (16).

the paragraphs following on from it. First was located the point on the hand, after which there appeared clauses which were as if its development, that is concerning the fingers. Further we have – in accordance with the content of the 1st point – the foot and eye. While in points 12 and 13 the order has been changed – for the nose is placed before the ear. Possibly this was dictated by the value of the compensation for damage to the eye and the severing of the nose – 45 and 15 *solidi*. In this case there has been adopted an ordering on the basis of the rate of compensation. In point 15 and 16 there is mentioned damage to the tongue and the knocking out of a tooth. It seems that the placing of these after the eye, nose and ear constitutes the next development of the list from the point of the former. All the parts of the body mentioned in paragraphs 12 to 16 are located in the area of the face – creating its own form of unity. In the points 17 and 18 talk is in turn on damage to the male genitals. This is the part of the body that is located on this list without connection to the content of the first point.

The set of bodily members enumerated in the first paragraph is not accidental. It appears that Frankish codifiers considered the hand, foot, eye, ear and nose to be among the most important parts of the human body. Therefore they also listed them in first place, allocating an extremely high rate of fine, 100 solidi, for their amputation. In this context attention is drawn by the fact that in the title *De uulneribus* there is first mentioned the skull (§§ 4–5), and then the torso (§§ 6–7). For if one were to compile the titles XVII and XXIX, with the missing out of title XXIX, the parts of the body would arrange themselves in the following order: skull, torso, hand, fingers, foot. One may interpret this arrangement as evidence of applying the ordering of the mentioned body parts from head to foot. Of course this model was used in an inconsistent way, and first and foremostly it did not constitute the single ordering scheme of things. Besides this crimes against the body were arranged according to the type of damage – wounding or the causing of disability.

The third criterion for the arrangement of regulation order in title XXIX was the value of the fines for particular bodily damage. It follows, however, to state that none of these methods of systemization were consistently applied. For example, the breaking of the skull bone or the wounding of intestines do not really accede to the remaining damages contained in the title *De debilitatibus*. In turn, in the title XXIX complete castration, for which the damages were 200 *solidi*, was placed in point 18 although paragraph 1 contained damages for which it was necessary to pay 100 *solidi*.

The Burgundian codification in relation to the place occupied in it by regulations on crimes against the body, is, against the backcloth of barbarian legal codes, a specific case. In almost all of the other *leges* (with the exception of the law of the Salic Franks) these regulations created a tight, often clearly isolated.

part of the text. While in Burgundian law they are scattered around various parts of the code, and first and foremost in comparison to other *leges*, their content is unusually selective. There is also not visible a criterion for the selection of these and not other types of crime and violations or damages to the body.

The assembly of regulations on crimes against the body, numbering in total 34 paragraphs, appears already on first glance fairly chaotic and most selective.[208] The regulations talking of freemen, the most important for our research, numbered a mere 13 paragraphs. These regulations give the impression that the regulations were created *ad hoc*, without a clear plan adopted from above or an organizing scheme. They cover cases of damage for only some of the parts of the body, and pass over the vast majority of the others.[209] Puzzling is the absence of many fragments of the body, which in other barbarian codes constituted a permanent component (e.g. the eye, ear, nose, male genitals, hand or fingers), with the simultaneous existence of regulations concerning those which appeared in them rarely – e.g. teeth. One may state with a large degree of likelihood that the Burgundian kings – Gundobad and Sigismund, as well as their advisers, did not strive to create a complete and systematic collection of rules regulating cases of violation against the human body.

* * *

The above analyses show that there existed within the researched barbarian codifications different ways of introducing regulations on crimes against bodily integrality and inviolability. We have shown several varied textual solutions, by means of which the said regulations have been ordered (and particularly the ordering of their enumeration) in the individual law codes. There remains the question – in what way should these practices be interpreted, and in particular – what do the differences between them point to? This question is connected with the problem of how we understand the human body, and in this case the collection of bodily parts. In particular the matter concerns the establishment of whether the said arrangements of regulations on the body were connected with a definite way (or various ways) of perceiving the bodily shell of a man or whether their occurrence and character could have been conditioned by other factors.

---

208 Regulations of this type are place in 8 different titles: *LConst*, V, 1–7, XI, 1–2, XXVI, 1–5, XXX, 1–2, XXXII, 1–4, XXXIII, 1–5, XLVIII, 1–4, XCII 1–4, XCIII. However, one needs to add that these titles contained regulations relating to various social groups; the regulations talking of freemen, the most important for our research, numbered a mere 13 paragraphs.

209 These are undefined parts on the body, the whole body: hair, arm, face, hand, teeth and shin.

Here arise two hypotheses. The first of them talk of the said textual arrangements that resulted from the means by which the given code arose or from the general compositional model of the given code adopted by its editors. These would, therefore, be certain editorial solutions which were created in a more or less unintentional way, and in any case without the clear intention of bestowing on the code of regulations of interest to us an definite arrangement and order. One should not discern in the said compositional forms any other message besides that which results from the content of individual regulations. The differences between the various *leges* in relation to the compositional differences within the groups of regulations in question would have, in essence, equally a secondary meaning. The second hypothesis is based on the premise that the said differentiating ways of ordering the collections of regulations on damages and violations could constitute the expression of certain cultural ideas functioning not only within the language of legal norms but enrooted in the mind set of Germanic societies of the era under research. There results, consequently, the question as to whether we are dealing, in the case of the main compositional solutions which result from our analyses, with the effect of different models – e.g. of traditional Germanic culture connected with oral transfer and the remains of a pagan religion, or whether equally of the Roman-Christian culture based on the written word. According to this conception, reconstructed compositional models would associate themselves with these two cultural circles on the basis of disconnectability.

In the light of the analyses presented one cannot unequivocally state that the order of enumerating crimes against the body according to the model of man's figure seen from head to foot was a manifestation of traditional Germanic culture or equally that it was derived from the Roman-Christian. This is shown by the fact that in a part of the codes which have the character of being of traditional laws transferred orally (e.g. Salic Law) there appears a different arrangement of the regulations. This does not mean, however, that in this case its source would be the influence of written Roman-Christian culture. Though admittedly there exist certain traces allowing one to suppose that Roman influences could have resulted in the use of the scheme of 'from head to foot'.[210] There are, how-

---

210 An example could be here the above cited letter of Sidonius Apollinaris to Agricola, containing an extensive and extremely detailed description of the physical appearance of the Visigoth king Theodric I; see chapter II, footnote 3. The description of the figure covers 37 parts of the body or bodily 'objects', namely: the figure (*forma*), the body (*corpus*), the skull (*apex capitis*), the forehead (*frons*), the occiput/back of the head (*vertix*), hair (*caeraries*), neck (*cervix*), eyebrows (*supercilia*), eyes (*geminos orbes*), eyelids (*cilia*), eyelashes (*palpebrae*), cheeks (*malas? maxilas*), ears (*aures*), nose (*nasus*), lips (*labra*), mouth (*os*), hairs (*pilis*), nostrils (*narium*), stubble (*barba*), cheeks

ever, serious arguments indicating that this actually caused a reluctance towards using any ordering model, as was the case in the Burgunian *Liber Constitutionum*. It is not though clear why the 'head to foot' scheme does not appear in Salic law (*Pactus* and *Lex*), Bavarian, or in *Lex Thuringorum*, where one can at best perceive only certain elements.

In relation to the above ascertainments it is possible to state that there did not exist a widely accepted single model within Germanic laws defining the arrangement of the regulations on crimes against the body and particularly the order in which they were to be written down. The arrangement of the regulations of interest to us, on the basis of 'from head to foot', although visible in many of the analysed codes, in some completely while in others partially, did not arise in an unavoidable way. It was the result of the choice of a specific means of creating this part of the code's text. One cannot, however, claim upon this basis that the adoption of another option (like, for example, in the case of Salic Law) had necessarily to result from some fundamental difference in the conceptualisation of the body, or more broadly from a cultural differentiation amongst the Franks, the Bavarians or the Thuringians, in relation to other Germanic peoples.[211]

(*gene*), chin (*mentum*), throat (*guttur*), neck (*collum*), shoulders (*umeri*), arm/shoulder (*lacerti*), forearms (*brachia*), hands (*manus*), chest (*pectus*), belly (*alvus*), back (*dorsum*), ribs (*costae*), spine (*spina*), torso/trunk (*latus*), thigh (*femur*), knee (*poplites, genus*), calves (*crura*), foot (*pes*). The author characterises the figure of the ruler, describing, point by point, the subsequent parts or fragments of the body, in an order from the top of his head to his feet. Similar, although not as detailed, descriptions exist in Roman literature much earlier on, an example of which being *The Twelve Caesars* by Gaius Suetonius Tranquillus. Sidionius' list, however, is of special importance from our point of view, for it concerns a Germanic leader and most importantly came into being at a time when there already existed the Gothic state in which several decades later *Codex Euricianus* evolved, considered to be the oldest list of laws issued by a Germanic ruler and the first Visigoth codification.

211 Cf. L. Oliver, *Beginnings...*, pp. 36 ff., referring to the laws of Ætherlberht, the author considers the regulations concerning bodily damage, to which was bestowed an arrangement from head to foot, represents a similar mnemonic scheme to that which organises the composition of the entire code (i.e. according to the social hierarchy from king to slaves/serfs). Referring to the thinking of M. Carruthers (*The Book of Memory: A Study of Memory in Medieval Culture*, Cambridge 1990, pp. 71–79) she claims that the arrangement of the group of regulations of interest to us is a manifestation of 'architectonic mnemonics' involving the creation of oral utterances according to a model supplied by some object – in this case the human body. Oliver sees in an arrangement of this type expression of the oral mechanism of creating law regulations. She rejects also the view that the arrangement 'from head to foot' is a representation of 'a logical and natural' way of ordering a group of regulations concerning crimes against the body for – as she correctly considers – this order was not applied in all the legal codes of ancient

The problem of compositional schemes should be examined within the context of the ways of understanding the concept of the body, and also the perception of relations between this concept and the collection of members or other bodily elements. As we remember, the laws of the Visigoths and Lombards differed from other barbarian codes in this matter, which resulted presumably from the influences of Roman-Christian written culture. However, it follows to add that the collection of regulations on crimes against the body in *Liber Iudiciorum* (partially) and in *Edictum Rothari* (fully) are examples of the use of the scheme 'from head to foot'. This same arrangement appears, however, in the Laws of Æthelberht and in the Pact of Alamann law, which does not display the influences of Roman law, while the elements of Christian origin are here not numerous.

In attempting to summarise the above considerations which have not led to the formulation of unequivocal theses, we shall look at another potential way of solving the problem. The *leges* herein analysed, although differing amongst themselves in relation to the means of textual depiction of matters of crimes against the body, do display certain common features. The most important of these is the widespread adoption of the principle (equally in *Liber Constitutionum*) that the law should regulate separately each case of violation of the integrality, more or less precisely of the indicated part of the body or other element of the human organism. This is a specific feature of the barbarian concept of criminal law, which resulted for sure from more general notions of man as well as of personal goods: of inviolability and bodily integrality, of health and life. This led to the creation of a collection of regulations corresponding to a longer or shorter list of the parts of the body that constituted the subject of legal regulations.

The Factors Forming the Way of Presenting the Structure of the Human Body in the *Leges*

Considerations on possible sources of understanding the concept of 'the body' in Visigoth and Lombard laws direct our thinking towards a more general question – here the matter concerns those factors which define the way and perception of members of the body in the Germanic legal codes. For here there appears the

---

peoples (e.g. *Pactus legis Salicae*, XVII, the Old Frisian code *Broker*, 177–204; Hittite law, table I, 7–17, *Code of Hammurabi* 196–209). The author concludes: 'Surprisingly, the head-to-toe layout occurs in none of these texts; clearly, albeit unexpectedly, this progression is not typical when laws are committed to writing.' (ibidem, p. 37).

question: what brought about a situation whereby the said texts, in talking of crimes against the body, mentioned so many individual parts of the body? One may differentiated at least three elements of this type:

1) the specific subject of interest and the regulative functions of legal texts,
2) the forms of information transfer specific for a culture of the spoken word and that of the written word,
3) the 'world-view' of the creators of the researched codes, developing from religious convictions and particularly the conception of man appearing within them, his temporal existence and that of the afterlife.

The role and scope of interaction of these factors in the forming of the concept of body and the perception of man's corporality require a searching analysis of the barbarian *leges* of interest to us. In this research the starting point will be the following hypotheses on the factors mentioned: firstly, the Germanic *leges*, despite the fact that they are available to us in a written form, display many features of information transfer and style of thinking that reflect spoken language (so-called oral culture); secondly, although they came into being in a period when individual Germanic peoples were already more or less Christian, clear traces of pagan convictions are present within them; thirdly, the legal norms of the barbarian codes were a part of the culture of the peoples under research, but their specific subject and function could have caused that the way of utilization and possibly also the meaning of individual concepts therein contained differed from those occurring in other forms of textual transfer. The three factors herein mentioned as well as the traits of the *leges* we have assumed to be present had an influence on the entirety of their content. For the problem under consideration fundamental is, however, their interaction on the understanding and means of utilising the concept of body.

These assumptions must, of course, undergo verification. Hence there is a need to examine the barbarian codes not merely in the context of legal norms (also those of the late Roman period) but also within the context of other sources expressing a different cultural content or presenting the question of the human body differently. Consequently indispensible is a confrontation of the normative sources with the narrative ones, of the forms and content of oral cultural transfer with those that are typical for the culture of the written word, and also the concept of man and his corporality developing from pagan religious beliefs with the ideas of Christian culture.

Grasping the specifics in the way of perceiving and presenting the body in Germanic laws, as has been observed in the hitherto analyses, is not an easy task when we view it merely on the basis of normative accounts. The incorporation in our investigations of other texts will not only highlight the distinctiveness in understanding the category researched in various types of textual pieces of evidence but also will, we conjecture, result – on the basis of contrast – in a reconstruction of those aspects of viewing the problem of the body, which are difficult to perceive from research only into the texts of the *leges*. Besides, thanks to comparative research the subject appears within a wider perspective and in a multidimensional way.

Concrete matters emerge from the above comments, ones that are to become the subject of our investigations. This first is the body of a man as the subject of legal norms, and more exactly – of the regulations on crimes against bodily inviolability, health and the life of an individual. The second problem is the influence of an oral means of knowledge transfer on the way the human body is perceived. This is connected with the complex matter of oral and written forms of transfer in the analysed legal codes. The next question is the way of comprehension and the utilisation of this category in narrative accounts. Finally, there will be examined the concept of perceiving the body within the frameworks of a pagan 'world outlook' on the one hand and from a Christian perspective on the other.

Barbarian *Leges* – A Type of Utterance on Man's Body as the Object of Crime

Investigations into the above mentioned factors determining the way in which the body is perceived and conceptualised should commence with underlining the fact that the researched legal codes constitute a specific type of utterance both with regard to the unique type of interest in reality as equally the functions that they fulfil. The attention of the barbarian codifiers was directed to the various crimes against the integrity of the substance of the body as well as man's bodily inviolability. This feature of the texts discussed has a significant meaning for our research. The regulations contained in the *leges* refers to concrete, strictly defined acts involving violation of the human body. Every clause described a separate case, and also concerned another fragment of bodily substance. Such a way of formulating the said regulations fundamentally influenced their content.

Indispensible, in explaining this question, is an attempt at recreating the means by which an individual clause was edited and the train of reasoning adopted by the codifiers. With this aim in mind we shall refer to a concrete case illustrating a typical means for barbarian law in formulating a regulation on an act violating the body of an individual. Here is the example: *Si quis alterum in caput ita plagauerit, ut cerebrum appareat et ei fuerit adprobatum, mallobergo chicsiofrit sunt, denarios DC qui faciunt solidos XV culpabilis iudicetur.* (If someone injures someone else in the head so that the brain is visible and it is proven, then according to the judicial formula *chicsiofrit*, a 600 denar fine or one of 15 *solidi* is adjudged).[212]

We are here dealing with a precisely described case of violation of a definite part of the body. This regulation at the same time regulates the presented situation in relation to the legal aspect – the way a court would have acted in the given case. The codifiers were interested in a concrete act (damage to the skull) and its effect (the displaying of the brain). The starting point was therefore the defining of the crime (the type of bodily damage). The cause for the creation of such a regulation was the intention of founding a legal regulation for the defining of procedure in the case of committing acts of this kind, known to its creators from experience. The logic of creating this type of regulation – one adopted by the barbarian codifiers – required a narrowing of the description of crimes to a single part of the body, to a single 'object'. The body as a whole was here of no significance. Other forms of damage had no significance either, which could have come about simultaneously with that described in the cited regulation. Such an approach on the part of the codifiers brought about the need to create an array of separate regulations for individual crimes. Neither the concept of the body nor its parts were necessary to them in this process. They were involved in each case with a concrete instance of violation of the body.

The main, if not the only function of this type of rules was the regulating of a situation that had arisen as a result of a crime. This function defined to a decisive degree the perception and presentation of the body as well as of its individual fragments. They were categorised as objects of crime. The context in which the said members appear in the *leges* was that of varied forms of physical violence. Other contexts – for example – the canon of beauty, care over hygiene, aesthetic practices, sensual delights, the language of gestures, to mention just a few possible ones – had here no application whatsoever. They were not connected with the regulative function of the law. Indeed, legal norms as a type of utterance had a designated practical aim. A description of the bodily structure of man was certainly not one of them.

The human body, or rather its individual parts outlined above, appeared in barbarian codes as the object of criminal activities of various kinds. The most striking feature in the description of these acts was the fragmentary perception of the body: each of the smallest parts was viewed separately. This 'fragmentary' means of viewing the human body emerging from barbarian laws is for certain connected with the structure of the regulations comprising them. For the mentality of the Germanic peoples, for example, the cutting off of an arm and

---

212 *PLSal*, XVII, 4.

the piercing of an arm constituted two different cases, which created two separate legal situations. Therefore it was necessary to have a separate description for each of these crimes. Noting them down in another way would presumably have resulted in the law being incomprehensible. For the barbarian mentality viewed the individual members separately, and also the individual crimes against the body. The combining of these two groups of elements resulted in a multiple number of individual crime cases.

The specific, fragmentary perception and presentation of the structure of the human body within the regulations examined meant that each part of the body was treated separately and without connection to other parts. This practice was not, however, an unavoidable necessity, but the effect of the functioning of a specific cultural system based on the existence of certain means of perceiving reality by legislators as well as the specific form of their communication. It seems, therefore, that although the normative character of the *leges* had a significant influence on the creation of statements on the body, they may not be treated as the only factor shaping their content. For it does not explain all the means of formulating it that we are to deal with in the texts of the *leges*.

Barbarian *Leges* and Forms of Thought Transfer in an Oral Culture

One of the means of explaining the specifics of the image of the human body as we can observe it in the Germanic *leges* is the hypothesis that it constituted a product of oral culture. In the Visigoth and Lombard laws there appears the word *corpus* in the understanding of the body of a living man, which constitutes a complex whole. Such an understanding of the concept 'body' (which was not used in other Germanic codifications) belonged, according to our conception, to the conceptual stock of the culture of the written word. This constituted, in our opinion, an element alien to oral culture.[213] The above hypotheses obviously require presentation and justification. For this purpose we shall first attempt to establish whether the regulations on crimes against the body contained in the Germanic *leges* were actually an expression of a way of thinking specific to an oral culture.

In answering the question as to the way in which the regulations of interest to us came about, we shall begin from an analysis of the type of utterance they represent. The most striking feature of regulations on crimes against the body appearing in the individual codes is that each of its fragments was regulated by a separate rule. Creating the said regulations involved the enumeration of the sub-

---

213 The concept of 'oral culture' I understand in the meaning advanced by W. Ong, *Orality and Literacy. The Thechnologizing of the Word*, London-New York 1982, pp. 31 ff.

sequent violations of various parts of the body. The list of cases of this kind was increased up to the moment when all the cases considered worthy of notation had been described. Each type of violation and each part of the body was treated separately.

These regulations were not narratively linked to each other – each constituted a separate whole. The assemblance of norms of this nature seems to be typical for oral thinking. For the utterance – in the culture of the oral word – involved, among other things, the adding of subsequent elements which created not so much the entirety as an entirety sequence. The collections of barbarian *leges* regulations of interest to us arose through the accumulation of regulations on individual crimes against the body. The subject of these regulations was not the body as a whole, treated as an object conceptually distinct, but the concrete members of the body viewed individually.

Oral thinking does not use abstract concepts and also does not create their definitions. Instead it relates itself to concrete objects (e.g. the arm) and situations (e.g. the cutting off of a given part of the body). The feature of an oral mentality is situational thinking. An utterance within a culture of the spoken word presents the undertaking of concrete actions in a way that is in accordance with everyday experience, concentrating on their practical aspect.[214] It appears that it is this specific case that we are dealing with in the barbarian *leges*. For the regulations concerning crimes against the body contain within them talk of certain typical repeated events and at the same time ones accessible for ordinary experience. The attention of the codifiers was directed towards the registration of cases in their existential context. Hence, presumably, the subsequent crimes were presented separately.

The feature of the regulations of interest to us, which often is connected to the above described additive structure of their set, is the repetition of certain elements. The majority of the regulations were edited according to one or two formulas.[215] In certain legal codes the regulations on the herein researched crimes started from the expression 'Si quis alium...' ('If someone someone else...'), later on there occurred a description of the act, while this would conclude with the expression 'componat solidos...' ('will pay X *solidi*..."). Formulas of this type repeat themselves a dozen or so, even several dozen times in each of the codes studied. This results for sure from the above described way of perceiv-

---

214 Cf. W. Ong, *op. cit.*, pp. 42 ff. The author states, among other thing, 'In the absence of elabotated analytic categories that depend on writing to structure knowledge at a distance from lived experience, oral culture must conceptualize and verbalize all their knowledge with more or less close reference to the human lifeworld, assimilating the alien, objective word to the more immediate, familiar interactions of human being.'
215 *Ibidem*, pp. 34 and 37 ff.

ing crimes against the body and the perception of the body itself. The said repetition of formulas gave, however, the researched textual fragments a specific rhythm. This is equally a feature of oral utterance.

There is repeated in the researched codes not only formulas but also words appearing in the descriptions of particular crimes. In *Edictum Rothari*, for example, the word *excusserit* (i.e., 'he cut off', 'he knocked out') appears 13 times in 30 regulations. While in *Lex Frisionum* the word *absciderit* ('he cut off') appears 10 times, *vulneraverit* ('he wounded') 6 times (for 89 regulations). Repetitions also concern the names of body parts. In Frisian law the word *brachium* appears six times, while *cubitum* fivefold. Similar occurrences are to be observed in the other Germanic law codes as well. This proves not only the lack of systemization in the regulations according to the criteria of parts of the body or types of crime but also the typical tendency for an oral mentality to repeat the trains of an utterance, which had already appeared earlier in it.[216]

In order to illustrate the full specifics of the 'composition' of the collective regulations on the crimes herein investigated we shall make use of the fragment of *Edictum Rothari* on the cutting off of fingers:

> 63. On fingers. If someone cuts off the thumb of another he shall pay a sixth part of the price which is estimated for the man if he had killed him.
>
> 64. On the second finger [i.e. the index finger]. If someone cuts off the second finger of another he shall pay 16 *solidi*.
>
> 65. On the third finger [i.e. the middle finger]. If someone cuts off the third finger of another, i.e. the middle one, he shall pay 5 *solidi*.
>
> 66. On the fourth finger [i.e. the ring finger]. If someone cuts off the fourth finger of another, he shall pay 8 *solidi*.
>
> 67. On the fifth finger [i.e. the little finger]. If someone cuts off the fifth finger of another, he shall pay 16 *solidi*.[217]

In this short fragment encompassing merely a few titles there is employed an almost identical formulation. However, the most striking feature is that all the regulations relate to the self same damage to the body. Besides all speak of subsequent fingers. The word *excusserit* is used here five times, while the word *dig-*

---

216 *Ibidem*, p. 39.
217 ERot, 63. *De digita manus. Si quis alii policem de manu excusserit, sexta patrem pretii ipsius, quod homo ipse adpretiatus fuerit, si eum occidissit, ei conponat.* 64. *De secundum digitum. Si quis alii secundum digitum demanu excusserit, conponat solidos sedicem.* 65. *De tertium digitum. Si quis alii tertium digitum de manu excusserit, quod est medianus, conponat solidos quinque.* 66. *De quartum digitum. Si quis quartum digitum excusserit conponat solidos octo.* 67. *De quinto digito. Si quis alii quintum digitum excusserit, conponat solidos sedicim.*

itus fourfold. For a way of thinking that had deeply assimilated the culture of the written word it is surprising that these regulations in referring to the cutting off of fingers should be enumerated in five separate points.

As can be seen the Lombard codifiers (and in general 'the barbarian') felt the need to repeat the same formulas and words. In the oral lecture of knowledge which we are here dealing with there was no room for generalisation and subordination.[218] Important, however, was the describing of individual cases of harm. In the same time the descriptions of subsequent acts resulting in bodily harm were formulated so as they would refer to actual events of this type. They were close to everyday experience. Therefore, there is no talk here of damages to the body or the cutting off of fingers in general, but each time about a specific crime against a concrete member.

In our reasoning we have intentionally used the example from the *Edictum Rothari* in order to show that despite the appearance within it of the body as an abstract concept of crime (*lesio corporis*), clearly visible are the 'oral' means of presenting its damage. We observe this same phenomenon in the Visigoth laws, where the influences of written culture were even more strongly present.

* * *

The oral character of the utterances within the barbarian *leges* as well as the influence of the thinking generating them on the way in which the body was perceived is not easy to show and define when the subject of the analysis constitutes only the legal texts herein evaluated. Therefore, it seems justified to refer to a source of a completely different character in order to show, upon this example, the variable way of expressing very similar phenomena and a different conception of the body. Those statements on the body contained in the *leges* we shall confront with an account shaped by the culture of the written word.

In discussing the problem of understanding the body within written culture we need to refer to an example which although contained in a narrative text, is descriptive in character – this is the literary portrait of Charlemage included in his *Life* by Einhard. We shall quote the said fragment *in extenso*, in order to analyse its constructional features and the way in which it employs the concept of the body:

> Charles was heavy and well built in frame, tall, but not extreme in proportion – for he measured seven of his own feet – his skull was round, eyes large and alive, his nose somewhat larger than average, beautiful grey hair, a happy and cheerful face. Hence his posture was exceptionally majestic whether he stood or sat; although his neck appeared to be rather too thick and short and his belly somewhat prominent, this was covered by the evenness of the rest of the members. He had a strong gait

---

218 Cf. W. Ong, *op. cit.*, pp. 37 ff.

and in the whole posture of his body he was most masculine; only his voice, despite being resonant, did not correspond to the splendour of the body.[219]

The description of Charlemage's physical appearance contains compositional elements of various sorts. A lot of emphasis is placed on the structural categories, presenting the body as a whole – such as: height, body build, its magnificence. The very concept of the body (*corpus*) is in the said description a separated, independent category of a comprehensive character. Einhard incorporated in his description also individual body elements, chiefly those located in the region of the head: the skull, eyes, the nose, hair, face. These are descriptively characterised (e.g. the rounded skull) as equally being qualitatively presented (e.g. beautiful hair). The author draws attention to physical features (e.g. the nose of average size) as well as features connected with disposition (e.g. the happy and cheerful face, the eyes alive). Beside this he includes in his description also the neck and belly. He uses also qualitative designations, e.g. heavy and well-built in frame, tall. He also talks about the evenness of the components (members), which covered the imperfections of the neck and belly. The above characterisation contains not only the strictly physical features of the body, but also the way in which he utilised them e.g. the gait, which was in Charlemage 'heavy', the majestic stature both in a standing position as when sitting. Another element of the description are the aspects of corporality, such as: the voice – resonant but not befitting the majesty of the king's body.

The description of Charlemage's appearance is not analytical but synthetic in character. Einhard presented the body of Charlemage as a whole, characterising closely only selected parts of it. The body is not only the overriding category in relation to the component parts but also an autonomic structure in relation to Charlemage's character. The author shows through this the existence of a corre-

---

219 *Einhardi Vita Caroli Magni*, [in:] Fontes ad historiam regni Francorum aevi karolini illustrandam, pars I, ed. R. Rau, Berlin 1955, cap. 22, p. 192: *Corpore fuit amplo atque robusto, statura eminenti, quae tamen iustam non excederet – nam septem suorum pedum proceritatem eius constat habuisset mensuram – apice capitis rotundo, oculis praegrandibus ac vegetis, naso paululum mediocritatem excendenti, canitie pulchra, facie laete et hilari. Unde formae auctoritas ac dignitas tam standi quam sedenti plurima adquirebatur; quemquam cervix obessa et brevior venterque proiectior videretur, tamen haec ceterorum membrorum celabat aequalitas. Incessu firmo totaque corporis habitidine virili; voce clara quidem, sed quae minus corporis formae conveniret.*[...]; this description is modelled on the characteristics of the appearance of Roman emperors, cf. Suetonius, *De vita Caesarum*, ed. J. C. Rolfe, Cambridge MA 1998; cf. J. L. Nelson, *Did Charlemagne have a Private Life?*, [in:] *Writing Medieval Biography 750–1250*, ed. D. Bates, J. Crickand S. Hamilton, Woodbridge 2006, pp. 15–28, there being a review of new works on Charlemagne.

spondence between the bodily features (the physical) and the psychic. The biographer's intention was the presentation of the physical exterior of the ruler as an entirely possessing an individual expression. Such a description could be contrived by someone thinking in a literary way, representing a literary way of thinking and culture, employing abstract concepts, rules of logic, defined categories. It is therefore an example of viewing the human body as characteristic of high intellectual culture.

The above fragment from *The Life of Charlemagne* constitutes a specific case of utterance on the subject of the human body. For this is an element in the biography of an actual person, and at the same time an individual man. In this respect it differs fundamentally from the texts we have analysed earlier. For in the *leges* we are dealing with a highly unindividualised unit, with a specific person, one could say, the representative of a given social group. Equally the parts of the body that appear there are specific in character, devoid of any personalised features.

In comparing the style of human body transfer in the *leges* regulations and in the *Life of Charlemagne* it follows to consider the earlier presented practice of subordinating crimes according to the enumeration of the body parts from head to foot. For we should draw attention to the fact that such a scheme also appears in the *Life*. As we mentioned above, in each regulation there appeared, on the one hand, man (the crime victim), who replaced the concept of body, as well as one or more parts of the body. We have, therefore, every time a conceptual relation: man – part of the body which defines the statement on the man's body. This scheme was repeated in every subsequent regulation. The form of expressing the human body, which in our belief can be perceived in many codes of barbarian laws, was therefore created through the assemblage of elements subsequently added. A statement thus formed corresponded to a defined concept of the body. It was not perceived as the overriding structure, to which became subordinated component elements. In the description of Charlemage's appearance we are dealing with another type of statement about the body. Both the structure of the text as equally the conception of the body are here of a different type. The individual members were simply not enumerated there but were subordinated to the entire concept of the body. In the description of Charlemage's appearance we are dealing with an intentionally created description of the presented figure. In *leges*, however, the image of a man's body that is possible to reconstruct was the secondary effect of the application of the principle of ordering the collection of rules concerning the discussed crimes.

Another example of using the concept of the body understood as a separate subject of interest for the author in relation to the presented figure are the descriptions of the ascetic practices of the monks and hermits operating in the 6th century in Gaul, which are contained in Gregory of Tours *Histories*. The various means of mortification as a context within which the body appears elicited the specific aspects of corporality and created a specific image of the body. We shall investigate this mechanism of the revealing of the body on the basis of several examples.

We find within the bishop of Tours' work descriptions of the life of the clergy, in which their bodies are straightforwardly named. Gregory writes about a certain hermit from the vicinity of Bourges – Patroclus – a man '[...] of exceptional holiness and piety as well as of extraordinary temperance [...]'. Besides Gregory adds that 'he always placed a hair shirt on a bare body [*puro corpori*] [...]'.[220] In a similar way the bishop of Tours presents the conduct and personage of another hermit from the environs of Nice – Hospicius '[...] who really mortified his body [...]'. Besides: 'he wrapped his bare body in iron chains [*ad purum corpus*] and put onto this a hair shirt'.[221] In the cited examples the ascetic practices draw attention, involving the reduction in the menu as well as – which is of importance for our subject area – the fact that in the given short characteristic of the holy men there appears the concept of the body (*corpus*) of a living man comprehended as an entirety. At the same time this concept is understood as a separate category, autonomous in relation to the person. The body was the object of ascetic practice, which meant that in these descriptions there arose between mortifying man and his body the relation of subject – object. The body was simultaneously understood integrally and abstractly in these descriptions.

A somewhat different image of the effects of ascetic practice may be observed in the autobiographical story of the hermit Wulfilaich, active in the environs of Trier, which is quoted by Gregory. Wulfiliach, in wanting to convert the local inhabitants to Christianity (and in particular turn them away from the cult of pagan deity monuments) constructed a stake on which he stayed 'suffering

---

220 Gregorii episcopi Turonensis, *Historiarum Libri Decem*, [further: Greg. *Hist.*] ed. R. Buchner, vol. II, Berlin 1956, V, 10; (English translation: *History of the Franks, by Gregory, Bishop of Tours*, transl. by E. Brehaut, New York 1916) [...] *cilium semper puro adhibens corpori* [...]. He practiced ascesis – abstaining from all alcohol, not eating any meat dishes. The author informs us also that Patroclus' diet comprised only water sweetened with honey and bread dunked in water and seasoned with salt; on the role of the holy men and their miracles in *Histories*, cf. W. Goffart, *Narrators of Barbarian History: Jordanes, Gregory of Tours, Bede and Paul the Deacon*, Princeton 1988, p. 174 ff.; K. A. Mitchell, *Saints and Public Christianity in the 'Historiae' of Gregory of Tours*, [in:] *Religion, Culture and Society in the Middle Ages. Studies in Honour of Richard E. Sullivan*, ed. T. F. X. Noble, J. J. Contreni, Kalamazoo 1987, pp. 77–94.

221 Greg. *Hist.*, VI, 6: [...] *qui constrictus catenis ad purum corpus, indumento desuper cilicio* [...]; he ate only dry bread and a few dates, [...] *during the days of Lent he lived off the roots of Egyptian herbs* [...].

greatly' in winter due to the frosts. For he had 'uncovered feet'.[222] 'For when winter fell I was so plagued by the icy cold that as a result of the severe frost sometimes my toenails fell off, and on my beard there hung water frozen like the wax dripping from a candle'.[223] Therefore we are dealing with another convention of presenting ascetic practices. For Wulfilaich is not talking about suffering or damage to the body (the word *corpus* here does not appear at all), but about himself and his experiences. This can be seen in the statements 'I stood [on the stake] suffering much' or 'I was so plagued by the icy cold that as a result of the severe frost sometimes my toenails fell off '. The hermit lists, however, individual parts of the body: feet, nails and beard, treating them as one with he himself. In a later part of the story he adds another form of body mortification, that is the restricting of his menu to limited quantities of cabbage and water.

However, there is subsequently to appear a new theme within his story, in which the concept of the body is to change. Wulfilaich, in converting the local inhabitants through his incessant attitude, decided with their help to destroy the monument to Diana located in the area (presumably the image of the Celtic deity).[224] When the stone effigy was overturned and then smashed to pieces with iron hammers, the hermit went to take sustenance. Then, however, as Wulfilaich recounted:

> '[...] the whole of my body was covered from head to foot with malignant blisters so that there was no possibility to find a space free the width of a single finger. I went then to the church and I took my cloth off in front of the holy altar. For I had there a small bottle full of oil which I had taken from the Church of St. Martin. With my own hands I smeared all of my limbs and anon fell to sleep. I awoke around midnight, got up in order to pray and saw my body entirely clean, as if no sore had ever been on me'.[225]

Now we are dealing with the concept of the body as an entirety. The narrator uses the term *omnis corpus meus* – 'my whole body', which in addition is en-

---

222 Greg. *Hist.*, VIII, 15, p. 178: [...] *sine ullo pedum perstabam tegmine.*
223 Greg. *Hist.*, p. 178: *Itaque cum hiemis tempus solite advenisset, ita rigore glaciali urebar, ut ungues pedum meorum saepius vis rigoris excuteret et in barbis meis aqua gelu conexa candelarum more dependeret.*
224 Cf. the publisher's commentary, Greg., *Hist.*, p. 179.
225 Greg. *Hist.*, p. 180: [...] *ita omne corpus meum a vertice usque ad plantam pusulis malis repletum est, ut locus, quem unus digitus tegerit, vacuus invenire non possit. Ingressusque basilicam solus, denudavi me coram sancto altario. Habebam enim ibi ampullam cum oleo plenam, quam de sancti Martini basilicam detuleram; ex qua propriis minibus omnes artus perunxi, moxque sopori locatus sum. Expergefactus vero circa medium noctis, cum ad cursum reddiendum surgerem, ita corpus totum incolomen reperri, acsi nullum super me ulcus appariusset* [...].

hanced by the formula 'from head to foot' (*a vertice usque ad plantam*). In another sentence the author uses the term 'all parts' (*omnes artus*), the content of which, as it seems, conceptually corresponds to the phrase 'my whole body'. Therefore, here the body was an integral structure and consequently the collection of all the members vertically ordered, from head to foot. This impression of a comprehensive structure is emphasised in the statement that on Wulfilaich's body there could not be found a place without boils 'the width of a single finger'. The body gained in this way a surface dimension, being in this fragment a synonym for skin. The surface of the body covered with blisters or sores (*pusulis malis repletum est*), was subsequently completely cleansed (*corpus totum incolomen repperi*) as a result of application of the wondrous oil from the Church of St. Martin.

The body of a living man is presented by Gregory as an independent narrative element. The author saw the said object of description as a entire structure not as the collection of individual parts or fragments unconnected with each other. Such an understanding of the concept of the body is specific to a mentality formed through the literary culture of the written word. However, we know from the story of Wulfilaich that Gregory, in writing about his ascetic practices connected with his evangelizing mission, did not always refer to that concept, but related equally to individual fragments of the body (feet, nails, beard). It seems, however, that the story of Wulfilaich's activities (similarly to other fragments about holy men) bears witness to the fact that Gregory perceived the said 'objects' in the context of the body as its manifestation.

The Human Body in Christian Eschatology and the Pagan Beliefs of the Germanic Peoples

Our hitherto analyses of the usage of the words *corpus*, *hreo* and *hraiwa* in the *leges barbarorum* have indicated that the way in which they were understood was closely related to human death. These terms, as we noted, most often denote a corpse or dead body. When the provisions of the barbarian laws focus on the human body, it seems reasonable to regard this issue from the perspective of religious beliefs relating to human death. Available evidence clearly suggests that in the majority of Germanic law codes, especially in the earliest versions, beliefs and convictions regarding the posthumous fate of the body were shaped by ancient pagan beliefs and the traditional "worldview" associated with them.[226] The

---

226 Cf. R. Schmidt-Wiegand, *Spuren paganer Religiosität in frühmittelalterliche Rechtsquellen*, [in:] *Germanische Religionsgeschichte. Quellen und Quellenprobleme*, ed. H. Beck, D. Ellmers, K. Schier, Berlin–New York 1992, pp. 575–587; eadem, *Spuren pa-*

second important question closely related to the issue of corpses is the matter of man's passage from life to death and the relationship between the living and dead body.

Any attempt to present the above issues should begin with the statement that the barbarian *leges* only contain fragmentary knowledge relating to human death. This is mainly due to the normative nature of the content they relayed. Convictions of a religious nature were only expressed in the provisions of these codes when there was a need to clarify their regulatory function. It also appears that the manner of perceiving the passage from life to death, especially the corporeal aspect of this phenomenon in the traditional culture of the Germanic peoples, will become clearer when confronted with Christian conceptions. According to mediaeval Christian conceptions, the body and soul coexisted within a human being and these were perceived as two discrete and unequal spheres. The body is shell (*foris*) and the soul is the interior (*intus*).[227] According to Gregory the Great, the body was supposed to be the 'abominable clothes of the soul', while St Paul described it as the 'temple of the Holy Spirit'.[228] These assertions relate to man during his earthly existence.

The main focus of our inquiries, however, is the issue of the fates of the spirit and body at the moment of death. In order to present the Christian vision of dying, let's again invoke the entry contained in Gregory of Tours' *Historia Francorum*. The story of the spiritually ascetic life of Salvius, Bishop of Albi, will serve as a good example. Having abandoned secular service at the royal court, he went to a monastery, where he spent many years. After some time, when he was already in the prime of life, he became head of the monastery community after the abbot's death. He was practising strict asceticism at the time, as a result of which his health had deteriorated. 'And once – Gregory goes on to relate – he lay panting on his bed worn out by a high fever, and behold his cell was suddenly brightened by a great light and quivered. And he lifted his hands to heaven and breathed out his spirit (*spiritum exalavit*) while giving thanks. [...] the monks took the dead man's body (*corpus defuncti*), out [of the cell],

---

ganer Religiosität in den frühmittelalterlichen Leges, [in:] *Iconologia Sacra. Mythos, Bildkunst und Dichtung in der Religions- Sozialgeschichte Alteuropas*, red. H. Keller, N. Staubach, Berlin New York 1994, pp. 249–262; eadem, *Wargus. Eine Bezeichnung für den Unrechtstäter in ihrem wortgeschichtlichen Zusammenhang*, [in:] *Zum Grabfrevel in vor- und frühgeschichtlicher Zeit. Untersuchungen zu Grabraub und „haugbrot" in Mittel- und Nordeuropa*, ed. H. Jankuhn, H. Nehlsen, H. Roth, Göttingen 1978, pp. 188–196.

227 J.-C. Schmitt, *op. cit.*, p. 66; cf. J. Le Goff, *L'imaginaire médiéval: éssais*, Paris 1985, pp. 123-125.

228 J. Le Goff, N. Truong, *op. cit.*, p. 35.

washed and clothed it and placed it on a bier and spent the night in weeping and singing psalms. In the morning while preparations for the funeral went on the body began to move on the bier. And behold his cheeks regained color and, as if roused from a deep sleep, he stirred and opened his eyes and lifted his hands [...] He rose from the bier, feeling no harm from the painful experience he had suffered [...]'.[229] In a further part of his life story of St Salvius, the Bishop of Tours relates the circumstances of his 'second' and this time final death – 'And when by God's revelation, as I suppose, he recognized the time of his calling, he made himself a tomb and washed his body and clothed it; and thus always intent upon heaven he breathed out his blessed spirit'.[230]

In this history, so rich and diverse in content, let's pay attention to the way in which the author presents the wordly and supernatural fates of the holy ascetic's body. In both death scenes the following opposition of ideas appears: body – soul/spirit. In the first of these, however, the dying Salvius, who 'breathed out his spirit' (*spiritum exalavit*), is mentioned before 'the dead man's body' (*corpus defuncti*), i.e., the corpse. Afterwards, however, that dead body is miraculously restored to life. For we discover that *corpus movere coepit*. We are, therefore, dealing with a living body. The term *corpus defuncti* strongly resonates with various formulae often encountered in barbarian laws: *corpus defuncti hominis*, *corpus occisi hominis* or *corpus mortuus*.[231] However, Gregory often also uses the word *corpus* in the sense of a living person.

In the second scene, the body – soul opposition of ideas is clearer. The body is at the same time treated as an object of Salvius' actions, for he performs the deathbed ablution on his own bodily shell. Gregory does not, however, say that the holy man 'washed himself', but that – 'he washed his body' (*corpus abluit*). The will and conscious action of the saint are distinct with regard to his body. Most importantly, however, here is the fact that the spirit (*spirit*) is something completely separate which leaves the body. Death in this understanding entails the separation of bodily and spiritual elements. At the same time, however, the body understood as a complete entity is an organism that can be described using medical or physiological categories, as is clear from the passage speaking of the return of the monk's bodily functions (the colour returning to his cheeks, his eyes opening, his body moving). It is also a shell. This is clear from the passage that speaks of skin shedding.

In the *Historia Francorum* of Gregory of Tours we find passages speaking of human death. The moment of passage from life to death was presented there

---

229 Greg, *Hist*, VII, 1.
230 *Ibidem*.
231 Cf. *Captulatio de partibus Saxoniae*, VII; *PLSal*, LV, 1 and 4 and *LRib*, LXXXV, 1.

as the separation of the soul from the body. A dying person issued his final breath and all that was left was the body or corpse (*corpus*). In the examined barbarian *leges* the relationship between death and the body is presented differently. First and foremost, the provisions of the Germanic laws conceive of human death (most often the result of murder or injury) as a situation resulting in the revealing of the presence of a body as a discrete structure. The fundamental difference in relation to the Christian vision of death, however, is that in the *leges* a human being is transformed into a dead body. In this process there is no conceptually graspable transformation of a living body into a dead one. The fundamental question arises here of how Germanic societies perceived the phenomenon of dying, the passage from life to death. This issue can only be resolved by referring to knowledge from the history of religions or religious studies fields about the relationship beween the spiritual and corporeal element in the life of man, both before and after death.

Gerardus van der Leeuw in his already classic *Religion in Essence and Manifestation: A Study in Phenomenology* formulated the following thesis when referring to the issue of the concept of the soul and its role in human life:

> [...] the soul is not a mere part of man, but the whole man in his sacredness; [...] This soul as a discrete entity is linked to a specific substance. Rather than being bound up with one particular part of the human body, it is spread over all its parts [...] exactly as is the case with blood which penetrates the whole body, though certain body parts have more of it than others. [...] It is substance or matter. Of course, in a primitive manner of thinking that is not familiar with the dualism of body and soul, this does not denote materialism.[232]

Van der Leeuw goes on to speak of the relationship between the soul, which he understands as a force, and the body and its separate parts. 'The force of the soul', he claims, 'is bound up with nearly all parts of the body and also resides in anything that emerges from the body.' In his view, such conveyors of the soul include breath and blood as well as spittle, sweat and urine. His line of reasoning continues as follows: 'The so-called "organ souls", forces relating to separate body parts, are of a different nature'. These include: the head, heart, liver and eye.[233] 'As far as the soul's first structure, i.e., structure as substance, is concerned, the soul actually represents a separation principle, but does not separate substance from force and even more so, body from soul [...]. Death is of no small consequence for this kind of soul. A dead person is not a soul, *but an en-*

---

[232] G. van der Leeuw, *Religion in Essence and Manifestation: A Study in Phenomenology*, transl. by J. E. Turner, New York 1963, vol. I, p. 275-276.
[233] *Ibidem*, p. 278.

tire dead person [italics – P.T.]'.²³⁴ 'The dead person then', writes van der Leeuw in another part of his book, 'is no soul without a body, but another corporeality (...).'²³⁵

Similar views on the way the Germanic peoples understood death have been presented by Edward Potkowski, who mainly draws on Scandinavian sources. The author's assertions include: 'Old Germanic peoples failed to grasp the essence of death. According to their beliefs, death did not put an end to life. The deceased went on to live a life similar to that which he had led before death.'²³⁶ The author goes on to write: '[...] Death did not put an end to life and human activity on earth. There was no difference between life and death. The deceased continued to need food as much as a living person.'²³⁷ He also refers to a crucial issue for us, the dead body: 'The dead Helgi', he writes, 'appears just like he would if he had fallen in battle: his head and chest are covered in blood and he has ice cold hands. So wounds received during life also appear on the body of the deceased after death. They can also be healed after death.'²³⁸

It would appear that these beliefs were reflected in the funeral practices known from archaeological research conducted in areas that were inhabited by Franks. M. Rouche characterises Merovingian Age cemeteries as follows: 'The increasing recourse to burial ... tended to foster belief in the notion that the dead inhabit a word of their own. The rural cemetery recreated the endogamy of the village.'²³⁹ The dead were buried fully clothed and equipped with objects for everyday use, such as: tools, a weapon, various implements (a comb, epilation

---

234 *Ibidem*, p. 283.
235 *Ibidem*, p. 128.
236 E. Potkowski, *Eschatologia germańska*, Euhemer – Przegląd Religioznawczy, 1963, no. 1 (32), p. 26. The author evokes *The second Poem of Helgi Hundingsbani*, in which the titular protagonist was the slayer of Hunding (Helga kvida Hundingdbana II) from the *Poetic Edda*. After the death of Helgi, who was supposed to reside in Valhalla, he demanded that his foe, Hundingd, serve him. He also returned to his own grave, where he entertained his sweetheart, Sigrun; cf. idem, *Dziedzictwo wierzeń pogańskich w średniowiecznych Niemczech. Defuncti vivi*, Warszawa 1973.
237 *Ibidem*, p. 27.
238 Ibidem, p. 28. The beliefs were also not alien to the southern Germanic peoples; the author invokes one of the rulings of the book of penitence of Burchard of Worms (beg. 11th century), which referred to the custom of burying the deceased with ointment, in order for him to tend to his wounds after death (*Fecisti et consensisti, quod quidam faciunt homini occiso cum sepelitur? Dant ei in manum unguentum quoddam, quasi illo unguento post mortem vulnus sanari possit, et sic cum unguento sepeliunt.*)
239 M. Rouche, *The Early Middle Ages in the West*, [in:] *A History of Private Life*, vol. 1., *From Pagan Rome to Byzantium*, ed. P. Veyen, transl. by A. Goldhammer, Cambrigde, Mass. 1996, p. 505.

tweezers, a fire striker) and jewellery (for women).[240] Eating vessels and foodstuffs (meat, groats, hazelnuts) were also placed in graves. 'In short,' concludes Rouche, 'the corpse ate, fought and loved like a living person. Materially the life of the dead offered exact parallel to the life of the living.'[241] Ruth Schmidt-Wiegand raises the issue, when writing about pagan elements in the *leges*, of the genesis of the provisions relating to the desecration of graves and claims that the objects mentioned in the title of *Pact* LV in the Salic Law, i.e. the post and shelter for the dead (*siselaubia*) as well as the *ponticulus* or *porticulus*, were understood, according to beliefs regarding the dead, as the 'doors of the dead'. Their existence enabled the deceased to participate in the life of members of his own family.[242]

The concept of the body as an object separated from and antithetical to the soul does not appear in the Germanic *leges*. It would, therefore, appear that death was perceived differently to the way it is in the *Historia Francorum*. When discussing the issue of understanding the body of a living person, we contended, on the basis of entries contained in the *Pactus legis Salicae*, that a person was perceived as a 'living soul', by which we meant 'a living being'. This conception presented the spiritual and corporeal aspects of existing as a person as being inseparable. At the same time, it does not appear that the body was perceived as an autonomous being in this conception. Instead, it was a manifestation of the existence of that 'living being' or 'soul'. When speaking of the conception of body in the barbarian *leges*, the clauses relating to human death should be taken into account. What is at issue here is its influence on perceptions of the body or, more precisely – the corpse. For the question arises of what status the body was meant to have in the investigated codifications. The second important issue is what happened to that 'living soul' after death.

In order to investigate these issues, let's turn back to the provisions contained in the *Lex Baiwariorum*, which most fully discuss the issue of the corpse. We earlier contended that title XIX of this code (items 5 and 6) was used to resolve cases of damage to the corpse. The first of these items related to damage inflicted on a dead body by an arrow aimed at scavenging birds, while the se-

---

240 *Ibidem*.
241 *Ibidem*, p. 507
242 R. Schmidt-Wiegand, *Christentum und pagane Religiosität in Pactus und Lex Alamannorum*, [in:] *Die Alemannen und das Christentum*, hrsg. von S. Lorenz, B. Scholkmann, Leitfelden-Echtringen 2003, p. 122; cf. B. Arhenius, *Tür für Toten. Sach und Wortzeugnisse zur einer frühmittelalterlichen Gräbersitte in Schweden*, Frühmittelalterliche Studien, vol. 4 (1970), pp. 384–394; PLSal, LV, 3: *Si quis charistatione super hominem mortuum capulauerit, mallobergo manduale, aut si laue, quod est ponticulus, super hominem mortuum capulauerit, mallobergo chreoburgio* [...].

cond spoke of the removal of a head, hand, foot or ear from the dead person's corpse and causing the flow of blood. At the same time, attention is drawn to the similarity between the content of item 6 and the provisions relating to violations to the body of a living person. Title IV of the law of the Bavarians includes item 14, in which the following was recorded: *Si aurem alicui absciderit cum 20 solidis componat*; while it is stated in item 2 that: *Si in eum* [sc. *liberum*] *sanguinem fuderit* [...] *solido 1 et semis componat*, and in item 9: *Si quis libero* [...] *manum vel pedem tulerit, cum 40 solidis componat*. In this title, there is in fact no mention (for obvious reasons) of decapitation, but several regulations relating to skull injuries are recorded.

These observations suggest by way of conclusion that the fact of death failed to cause any fundamental change in the attitude of the authors of these provisions to bodily injuries. A dead body was subject to the same legal protection against (the very same) acts of aggression as the body of a living person. The significance of these provisions cannot, however, be explained in utilitarian terms or on the basis of the realities of earthly existence. Therefore, it appears that they need to be investigated with regard to religious beliefs, i.e., early pagan beliefs. The body was no less important after death and did not simply become mortal remains. It was still needed, which is why injuring it was a crime, whether this occurred during a person's lifetime or after death. Violating its integrity was harmful to an individual after death, for the body was still regarded as an aspect of a person as well as his 'property'. There was no radical difference between a living and dead body. During life, the invigorating (spiritual) element prevailed in a human being, while after death it was the body that came to the fore. This also gives rise to the impression that, in the second situation, the body manifested its existence in a 'concrete' form. Nevertheless, as we noted above, there was little essential difference between perceptions of the living body and the dead body.

* * *

Summarising the above deliberations, I would like to synthetically discuss the reasons for the concept of body not being related to a living person in the majority of the *leges barbarorum*, while using the word *corpus* in the sense of 'corpse' or 'dead person'. This phenomenon can be explained through close reference to three previously mentioned factors that helped to shape the content of the German law codes. The first factor is the focus of interest and the regulatory function of the *leges*. The object of 'discourse' of the investigated barbarian law provisions was physical aggression, in particular damage and injury inflicted on separate body parts. This theme led to the human body being perceived in terms of fragments. The function of the law in this case consisted in the resolution of situations arising as an outcome of specific acts. The concept of a living body does not appear in these, because the creators of these provisions were interested in cases of injury to particular body parts. So the category of body was not necessary for the codifiers. But the regulatory function of the law nevertheless contributed to the creation of an extensive list of separate body parts.

The second factor which can be regarded as a reason for this particular way of perceiving the body is the manner in which knowledge is created and transmitted in an oral culture. The fact that the concept of the body as a complex autonomous structure failed to appear in most of the Germanic law codes can be explained by the oral mentality of the authors of these provisions and the manner of formulating ideas that was associated with this mentality. This is evident in the laws of the Franks, Anglo-Saxons, Alamanns, Bavarians, Saxons, Turingians and Frisians. In these, the concept of body was replaced by the concept of person. They also contain no 'body part' category. There was no attempt in them to categorise either the body or its parts. They were basically a set of unconnected elements. Similarly to the narrative texts that arose from oral culture, the concept of body was not a defined, abstract category. The relationship between the body as a discrete entity and its separate parts was vague. The concept of body was not systemised.

The Visigoth and Lombardian codes, which were largely subject to Roman written culture, contained a more abstract conception of the body. It should be noted, however, that, even in these codifications, the provisions relating to crimes against the body were created separately for individual body parts in such a way that successive violations were treated as discrete cases. In the narrative texts generated by a written culture we are dealing with a defined and abstract concept of body. It was supposed to be an integrated entity – a system composed of interlinked parts (members). The body was conceptually distinct from the individual.

The third way of explaining our issue refers to the 'worldview' religious factor that shaped the *leges* texts. In the case of the body, it is best to explain the impact of this factor by referring to the circumstances after death. We are dealing in this case with a way of thinking about a dead body in terms of the traditional categories of Germanic (pagan) beliefs. This is indicated by the provisions contained in the *Lex Baiwariorum*. These mention harm to unburied bodies. The kinds of injuries are strongly reminiscent of those inflicted on living people. The codifiers created provisions relating to the harming of corpses, because they were convinced that injury after death was harmful to the injured party. It therefore appears that in the barbarian *leges* we are not dealing with the Christian concept of man as a spiritual-corporeal being with a dichotomous structure, either in the case of a corpse or a living body. They were conceived from a 'worldview'[243] whose source was the pagan beliefs of the early mediaeval Germanic peoples, superfically modifications by the influences of the then Christian 'anthropology'.

---

243 I am employing this category being fully aware that it is somewhat incongruent with the mentality of people of the discussed period.

Chapter Three

The Human Body as an Object of Crime. Body Parts, Their Damage and Violation and the Compensatory Tariff System.

In our earlier deliberations, our analyses mainly focused on the issue of the way in which the creators of the *leges* understood the concept of human body. We now intend to discuss another issue important from the point of view of the subject of this book, namely separate parts of the human body viewed as objects of crime, for our main outlook on the body in the *leges* regulations is provided by descriptions of damage and violation. This assertion compels us to pose a fundamental question: How did the regulations pertaining to these crimes define the scope and manner of presenting individual body parts? The manner and degree of bodily injury was in fact a textual mechanism thanks to which specific body parts became a focus of interest for the creators of the legal regulations contained in the barbarian *leges*. The types and scope of injury to an individual's bodily inviolability and integrity were therefore an important issue. The second fundamental issue associated with the body as an object of crime is the composition tariffs set down by those compiling the laws in cases of particular damage or violations to the body. In the surveyed material, descriptions of harm to the inviolability and integrity of the body are inseparably linked to those tariffs which crime perpetrators were obliged to pay out to crime victims. Since they are expressed in monetary terms, these sums fail to provide any direct information about the body or corporality, yet they are crucial for our research, because they enable us to make an attempt to reconstruct the methods that were used for assessing bodily injuries resulting from various types of crime, and also to 'valorise' the human body.

In the vast majority of Germanic codifications, a perusal of the regulations pertaining to crimes against the body makes the contemporary reader aware of a particular structure for presenting individual regulations, which is repeated in almost every paragraph devoted to acts of this type. A model clause of this kind contained three main components which were always in the following order: (1)

the name of a body part or object, (2) the kind of injury (and, if need be, what caused it), (3) the composition tariff. The way in which the content of these regulations is laid out would appear to offer an important indication of a method that could be used for researching the issue that is of interest to us. For, according to our hypothesis, it expressed the manner in which the body was perceived and presented as an object of crime. As the creators of the Germanic law books were formulating individual regulations, they based every clause on a particular body part. In our view, this assertion justifies our method of analysing the material under discussion, a method which is based on an initial presentation of sets of damaged body parts or body objects appearing both in individual codes and, for purposes of comparison, in the complete set of surveyed codifications. For in our research we wish to examine rules relating to the body within the context of the compositional structure of the groups of regulations in which they are located. This method enables us to explain the meaning of rules relating to the body and the manner in which those who created them intended them to be understood.

For the above reasons, we will deal with crimes that do harm to an individual's bodily integrity and inviolability at a later stage. We will investigate them as a textual factor defining the content of statements about various body fragments. We will also investigate these sources within the context of any bodily and/or moral harm which the crime victim suffered. The next issue – composition tariffs – is closely connected to this. According to the barbarian understanding of penal law, the loss or harm caused by a crime led to a need for the perpetrator to compensate the injured party with a specific sum of money (or something of equivalent value) suited to the nature of the deed. In our research we will be taking into account the particular logic of the sets of regulations under discussion here. This is also why we will deal with the issue of crime types before the composition values ascribed to individual cases. We will treat both of these issues as indicators creating contexts within which individual body objects appear.

Body Parts as Objects of Crime – an Analysis of the Content of Selected *Leges*

The sets of regulations pertaining to crimes against the body contained specific lists of body members subject to harm and injury. These catalogues were selective in nature. The specific choice of body parts recorded in each *lex* was surely made on the basis of the criteria adopted by the people editing the legal code in question. The number of these 'body objects' varied from list to list. The variation between individual *leges* is presented through a comparison of data from the

13 surveyed law books, with the list items being arranged in numerical order from the lowest to the highest. These data of course require interpretation. They only offer a general oversight of the prevailing trend in the barbarian *leges* under discussion. A healthy dose of scepticism is advisable when approaching them from a statistical perspective.

Table 1. *The Number of Body Parts as Objects of Crimes in Individual Germanic Leges*

Name of legal code	Number of body parts	Percentage of the number of regulations on crimes against the body in the entire legal code*(%)
Liber Constitutionum (Burgundians), beginning of the 6th century	7	3
Lex Ribuaria (Ripuarian Franks), beginning of the 7th century	12	12
Liber Iudiciorum (Visigoths), mid 7th century	18	3.5
Pactus legis Salicae, 1st half of the 6th century	19	6.5
Lex Salica, 2nd half of the 8th century	20	7.5
Lex Thuringorum, 802–803	19	35
Lex Baiwariorum, 2nd quarter of the 7th century	24	12
Edictum Rothari (Lombards), 643	25	8
Pactus legis Alamannorum, 1st half of the 7th century	26	32
Prawa Æthelberhta (Anglo-Saxons), beginning of the 7th century	33	43
Lex Saxonum, 802–803	33	20
Lex Alamannorum, beginning of the 8th century	40	22
Lex Frisionum, 802–803	60	45

A comparison of the percentage divisions contained within the regulations of interest to us show a significant difference between the legal codes of the Burgundians, the Visigoths, the Salic and Ripuarian Franks and the Lombards (from 3% to 8%) on the one hand and the codes of the Thuringii, Anglo-Saxons and Frisians (from 32% to 45%). Besides, there is visible also the following regularity: in the more developed codes (i.e., where there was a large number of points referring to individual legal questions) there appeared at the same time a small number of injured body parts, also usually in the less developed codes we are dealing with a large number of members (e.g. in *Leges Visigothorum* – 582 individual regulations and 18 body parts, while in *Lex Frisionum* – 196 and 60 respectively)

Despite these reservations, one can notice, basing onself on these data, certain features within the surveyed set of laws and its components. The most arresting property of this comparison set is probably the significant range of numbers it contains (from 7 to 60 elements). This assertion should be furnished with a basic commentary. The Burgundian law book is a special case in this respect. The manner of presenting crimes against the body is so different to that of the other codes (and not only with regard to the very small number of elements it contains) that any statistical comparisons would be of doubtful value. Yet if we focus our attention on another codification (Ripuarian Law), the difference in relation to Frisian law is also very significant (12 to 60). It is also possible to isolate two groups of codes from this comparison set in which the number of body objects is relatively close. The first of these includes: the *Liber Iudiciorum, Lex Thuringorum, Pactus legis Salicae, Lex Salica, Lex Baiwariorum* and *Edictum Rothari, Pactus legis Alamannorum* (from 18 to 26 items); and the second: *Æthelberh Laws, Lex Saxonum* and *Lex Alamannorum* (from 33 to 40 items). Compared to these, the *Lex Frisionum* is clearly an exceptional case.

The cited data only portrays the surface layer of the investigated phenomenon, i.e., its quantitative aspect. The crucial issue is to establish what body parts these regulations are referring to, for the surveyed lists of body parts vary not only in number but also in type. It is therefore necessary to cite data of a qualitative nature. What is of concern here are the lists of body objects reconstructed on the basis of the regulations recorded in separate barbarian *leges*.

Table 2. *The Elements of the human body constituting the objects of crimes on the base of the laws of Ripuarian Franks, Alamanns and Frisians*

Lex Ribuaria	Lex Alamannorum	Lex Frisionum
1. unspecified place on the body	1. unspecified place on the body	1. head
2. bone	2. skin of the head	2. skin of the head
3. torso	3. skull	3. skull
4. ear	4. ear	4. cerebral leptomeninx
5. nose	5. upper eyelid	5. ear

6. Eye	6. lower eyelid	6. nose
7. hand	7. eye	7. the highest line (wrinkle) on the forehead
8. thumb	8. nose	8. the second line (wrinkle) on the forehead
9. index finger	9. upper lip	9. the third line (wrinkle) on the forehead
10. foot	10. lower lip	10. eyebrow
11. toe	11. front teeth	11. upper eyelid
12. genitals	12. molars	12. lower eyelid
	13. tongue	13. moustache/whiskers
	14. face	14. cheek
	15. neck	15. front teeth
	16. hair	16. side teeth
	17. beard (stubble)	17. molars
	18. elbow	18. neck
	19. hand	19. rib
	20. arm bone	20. arm bone above the elbow
	21. upper part of the thumb	21. arm bone below the elbow
	22. whole thumb	22. both arm bones
	23. 1. part of an index finger	23. hand
	24. 2. part of an index finger	24. thumb
	25. whole index finger	25. index finger
	26. 1. part of the middle finger	26. middle finger
	27. 2. part of the middle finger	27. ring finger
	28. whole middle finger	28. little finger
	29. 1. part of the ring finger	29. the five digits of the hand
	30. 2. part of the ring finger	30. the hand (without fingers)
	31. the whole of the ring finger	31. 1. part of each of the 4 long digits
	32. little finger	32.2. part of each of the 4 long digits
	33. the side	33.3 part of each of the 4 long digits
	34. internals	34. wrist

		35. genitals	35. elbow
		36. legs	36. shoulder joint
		37. knee	37. 1. thumb joint
		38. shin	38. 2. thumb joint
		39. feet	39. 3 thumb joint
		40. hernia	40. eye
			41. chest
			42. pericardium
			43. the membrane between the liver and pancreas
			44. stomach
			45. intestine
			46. belly
			47. penis
			48. 1. testicle
			49. thigh
			50. shin
			51. foot
			52. big toe
			53. 2 toe
			54. 3 toe
			55. 4 toe
			56. 5 toe
			57. back
			58. lung
			59. scrotum

LRib, I–VI, *LAl*, LVII, 1–69, *LFris*, XXII, 1–89. The lists of body parts have been compiled on the basis of the contents of clauses relating to crimes against the body contained in the legal codes of the Ripuarian Franks, the Alamanns and the Frisians. This is not, however, a literal repeat of the data on body parts as contained in the relevant paragraphs. We have taken into consideration those of them where the given object of crime appears for the first time, hence we have passed over those where a secondary reference is made within the context of another type of crime. The numbers of the individual items are cardinal numbers and are not identical with the numeration of the paragraphs themselves.

The lists of body parts presented in Table 2 appearing in the *Lex Ribuaria*, *Lex Alamannorum* and *Lex Frisionum* reveal three different selection variants. The list reconstructed on the basis of the regulations of the law of the Ripuarian Franks is an example of a dramatically reduced set, while that from the text of the law of the Frisians is a particularly extensive version. These delineate the minimum and maximum range of the contents of those catalogues in the surveyed set of law indexes. This does not mean that we should treat them as exceptional cases, for every variant of these lists is original and unique, and that also applies to the law of the Alamanns. These lists may also be viewed differently - questions emerge as to their completeness or lack of completeness, and also the possibility of an objective or universal scheme for describing their contents. A comparison of the Ripuarian and Frisian laws shows that these catalogues could have been extended or increased in detail without noticeable limitations. It is not possible to speak of a complete model list as such. It is also not clear whether it is possible to isolate a list containing an essential minimum of elements.

The feature that strikes one first and foremost in the above listing is the recalled differentiation in the number and type of body part. In examining the content of the individual lists we shall attempt through a defining of the said differences to show their characteristic properties. The differences between the, individual codes of barbarian laws contained herein manifest themselves in a dual fashion. In certain codes *Lex Alamannorum*, *Lex Frisionum* and the *Æthelberht Laws*, there appear parts of the body which are not taken into consideration by other codes. In the law of the Frisians there is talk of, for example, the three lines on the forehead, the upper and lower eyelid, moustache/whiskers (*granones*), the neck, a fingerless hand (*palma manus*) as well as feet without toes.[244] While in the law of the Alamanns, for example, the tip of the nose, the neck, the beard (stubble).[245] In the Kentish code of Æthelberht there are mentioned: the lower jaw, the collarbone, the nail of the digits of the hand and the toenail of the big toe.[246]

The second factor differentiating the number of body parts was the practice adopted in certain codes whereby regulations were drawn up on the damage inflicted on smaller body part fragments which otherwise appear as wholes in the remaining *leges*, as well as the citing of detailed names for the individual body parts instead of using a general term. In order to visualise this general principle we shall refer to the example of a legal listing wherein the described phenome-

---

244 *LFris*, XXII, 11–13, 22, 34, 62.
245 *LAl*, LVII, 9, 22, 24.
246 *Æthel*, 50; 52, 1; 54, 1; 72.

non is not present. We shall show against this background the distinctiveness of the practices employed in other *leges*. In *Pactus legis Salicae* there is talk of damage to those parts of the body as: the fingers, teeth, innards, hand, foot.[247] This is an example of a collection of regulations in which there have been employed terms for body parts that are understood as whole elements and inseparable (e.g., the fingers) or internally undistinguished categories (e.g., innards, teeth).

While in the codifications with more elaborated cataloguing of injured body parts we can see beside the above mentioned members, a series of other bodily elements. A particularly clear example of this are the regulations contained in *Lex Alamannorum* and in *Lex Frisionum*, which talk of the cutting off of two or three sections of fingers.[248] In the majority of the codes there is mention only of injury to fingers in their entirety.[249] A similar case may be seen with regard to teeth. In Alamann law we read of upper and lower front teeth, of 'cheek' teeth (*marczan*) and others, while in that of the Frisians of front, back and molars.[250] In the remaining codes they are defined as a single general category. Another example are the innards and in particular the stomach cavity. The regulations appearing in Frisian law are in this regard especially detailed, with the appearance here of the following designations: pericardium (*praecordium*), the membrane dividing the spleen and liver (*mithridri*), the stomach, intestine, belly, stomach.[251] A similar matter occurs with the upper limbs. In the listing for Alamann law there is mentioned injury of the arm above and below the elbow, of the elbow itself, the hand, and the whole arm[252], while in *Lex Frisionum*: the arm above the elbow, the arm bone above the elbow, the hand, a fingerless hand, the wrist, elbow, and the whole arm.[253] In the majority of the remaining codes there is talk usually of the hand and arm. The same applies to the lower limbs.

A comparison of the regulations in *Pactus legis Salicae* with those appearing in *Lex Alamannorum* and *Lex Frisionum* shows that in the latter we are dealing with a practice involving the showing of smaller objects within the limits of the categories appearing in Salic Law and other barbarian codifications. There

---

247 *PLSal*, XVII, 6 (innards); XXIX, 1–3 and 10–11 (hand and foot), 4–9 (fingers), 16 (teeth).
248 *LAl*, LVII, 41–53, *LFris*., XXII, 28–42.
249 *LVisig*, VI, 4, 3 (s. 265); *PLSal*, XXIX, 4–9; *LSal*, (D) XLVIII, 3–7, (S) XVI, 4–10; *Æthel*, 54,1–5; *LRib*, V, 5–7; *ERot*, 63–67, 89–93 i 114–118; *LBaiw*, IV,11;V, 7;VI,7; *LThur*, XIX–XXI.
250 *LAl*., LVII, 20–25; *LFris*., XXII, 19–21.
251 *LFris*, XXII, 49, 52–57.
252 *LAl*, LVII, 31–33, 35–40.
253 *LFris*, XXII, 24–25, 27, 34, 38–39, 76–79.

does not result, however, from this that those codifying Alamann Law or that of the Frisians used the legal text of the Salic Franks or some other legal code, although obviously one may not entirely exclude such an undertaking. It appear, however, rather as if the individual *leges* were created first and foremost on the basis of the indigenous traditions of the given people. The examples of the differences amongst the cited barbarian *leges* herein given, with regard to the type and number of body parts, affect not only the quoted codifications but display a tendency characteristic for all the legal texts studied.

Despite the indicated differences amongst the listings of body parts within the three codes presented they show (in a way similar to the remaining barbarian *leges*) certain similarities. The common feature that most strikes one is that in each of the researched groupings of regulations we are dealing with the assembly of selected, individual 'points', which the creators of the said regulations viewed within the limit of a man's body. Such an arrangement of content resulted from the fact that each description of crimes referred to a single 'bodily object' or a single body part. Statements on the varied bodily damage resulted in the isolating of a certain number of individual objects of crime. This impression is enhanced by the fact that each regulation presented a description of a separate legal situation or equally an individual event. They constituted separate micro-narratives, which were not connected with one another.

Body Parts – an Overall Look at the *Leges* Collections

A comparison of the catalogues of body parts contained in Ripuarian, Alamann and Frisian law shows difference amongst them both with regard to the number as equally the type of bodily objects. Although the said variations are an obvious fact this does not, however, mean that within the set of the thirteen examined legal codes there are not common features when the matter concerns the aforesaid selection of body parts.

Therefore, we shall look for now at the lists of body parts contained in all the codes as a kind of collection, treating their content as a whole. Thanks to a comparison of them with each other we may determine the similarities amongst the collections of regulations of interest to us, that is, first and foremost, as a collection of those fragments of the body that repeat themselves in the definite majority of the codes. This allows for an understanding of the mechanisms for the creation of concrete classes of body parts, both simple as well as those more elaborated in structure.

A catalogue of the body parts appearing in the thirteen codes studied covers 70 items.[254] The content of the source material induces one to differentiate two groups of body parts – the first containing those items which appear in the vast majority of codes, and the second covering those that manifest themselves less often or sporadically. This method has as its goal not only the presentation of which body parts were a common part of the individual legal codes and which should be treated as a supplement to the said 'basic set' but also it serves to emphasise the specifics of the individual codes. An analysis of the frequency by which individual body parts appear in the researched law lists shows that an assembly of common elements covers 15 body parts, therefore a little more than one fifth of the items named in the catalogue.[255]

Applying the order of body division into sections (the head, the torso, the upper limbs, the lower limbs) we shall ascertain that in the region of the head there belong 4 body parts to the 'basic set': the eye (in 12 codes), the nose (in 11), the ear (in 11) as well as the skull (in 8). In the section of the torso the common features that need to be enumerated are the torso itself (in 9) as well as the penis (in 8). With regard to the upper limbs we have more body parts repeating themselves in the majority of the texts examined, these being: the thumb (12), index finger (12), middle finger (11), ring finger (11), little finger (11), hand (11 codes) and the whole arm (8). In the section of lower limbs one may

---

254 In this number there are found 24 parts of the body in the area of the head and face: the head as a whole, the skull (bone), the brain (and also the membrane lining the skull), hair, facial growth (whiskers and beard), the forehead (specifically: three transverse wrinkles/lines dissecting the forehead), eyebrows, eyelids (upper and lower), nose (nostrils), lips (upper and lower), teeth (front, side, molars, upper, lower), ears, tongue, cheeks, face (specifically: the skin where there was no hair growth). The next 17 body parts are situated in the area of the torso: the neck, throat, shoulders, collarbone, whole torso (back, chest, belly), ribs, innards (pericardium, stomach, intestines, the membrane between the lungs and the stomach), the genitals (the penis, testicles, scrotum). The next group are the parts comprising the upper limb (15 items); the whole arm, the arm above and below the elbow, the elbow, the wrist, hand, fingerless hand, thumb, index finger, middle finger, ring finger, little finger, the first part of a finger, the second part, the third part, the nail of each finger). The codes researched enumerate 12 parts of the body for the area of the lower limbs: leg, thigh, shin, knee, the fifth toe, the nail on each toe. Besides this one may note two other items: an unspecified bone and the skin on the whole body.
255 We shall here adopt the premise that if a given part appeared in at least 8 codes then it follows to incorporate it into the set of common features which we shall call for the purposes of our research the 'basic set'. While if we come across the given body part in less than seven texts we shall consider the said to be an additional element in relation to the 'basic set'. After applying this criterion our cataloguing will significantly change.

mention only the foot which appears in 11 codes. In addition there needs to be included in this 'basic set' undefined places on the skin (9).

The vast majority of body parts mentioned in our catalogue (i.e., around 75%) do not fulfil the criterion of frequency of appearance herein designated. Therefore, they are included within additional elements to the set. A significant part of these appeared in the researched legal codes quite sporadically (1-2 times). It follows to add, however, that a certain part of the items are situated in the group between the two above mentioned categories. It follows to include in this group: molars (7) and front teeth (7), the whole leg (6), any bone whatsoever (6), the tongue (6).

We have in the region of the head a series of body parts appearing in only a few codes, namely: both lips (5), the brain (3), both eyelids (3), three lines (wrinkles/furrows) on the forehead (1), a cheek (2), eyebrows (1). In the case of the torso region the said rarely or sporadically appearing body parts are: the stomach (4), innards generally (3), ribs (3), the intestines (1), the pericardium (1), the membrane between the liver and the spleen (1) the shoulder (1), throat (1), chest (1), vulva (1). If the matter concerns the upper limbs then we should include within additional elements: the arm above the elbow (5), the arm below the elbow (4), the first part of the finger (3), the second part of the finger (2), the third part of the finger (2), the wrist (1) as well as the elbow (1). Finally additional elements for the lower limbs include: the thigh (5), the whole leg (4), the five toes (4), the shin (3), a toeless foot.

* * *

In researching the frequency of appearance of the above listed body parts, we have equally defined their (small in number) group which appears in the majority or even in the overwhelming majority of the *leges* researched, as equally the numerically dominating 'mass' of those items which appear in them either rarely or quite sporadically. The designation according to the numerical criterion of a 'basic set' of the said bodily objects is able to give one merely a general concept about the decisions taken by those who devised a given legal code. A hypothesis that states that the group of these 15 most frequently occurring elements, being its own form of common part for the entire set studied, reflects something in the nature of a 'primary matrix' that constituted the basis for the creation of a list of crimes against the body within all the barbarian codes, although undoubtedly attractive, is difficult to accept without reservations. For how is one to explain the case of Ripuarian Law where 12 body parts appear? One may equally ask the question – why were none of the remaining codes etirely in agreement with this said 'primary matrix' or 'basic set'?

If we consider that the hypothesis of a 'primary matrix' does not entirely explain the mechanism of the creation of the said catalogues of body damage, there arises the question as to what factors decided on the choice of a given set in the particular cases. This is at the same time a part of a broader problem: what in general decided on the placing on the said lists of any fragment of the body whatsoever? As we have ascertained above we see the body of a man in the codes under investigation through the descriptions of its damage and violation. One may therefore advance the hypothesis that the factor deciding as to the contents of such a list of body parts as appearing in the given code was the varied forms of crime afflicted against it as well as the outcomes for the victims. There exists, however, another way of explaining this question. The catalogues of body parts appearing in the various *leges* were in their own way models, through the help of which the said crimes were presented. The lack of one 'primary matrix' does not mean that the individual codes did not contain their own schemes of this type.

An explanation of the problem of the genesis of the catalogues of body parts is not simple. The collections which were recorded in each of the codes under consideration were not of the character of a universal model automatically copied by subsequent creators of codifications of barbarian laws. The process of their creation was presumably subjected to practical necessities resulting from the specific function of the *leges*. Undoubtedly the selection of individual 'bodily objects' was defined by the fact as to whether their damage had influence on the life situation (health, social) of the victim. One may, therefore, assume that those body parts that were selected were significant in relation to the above consideration. This ascertainment, although not devoid of correctness, cannot be treated as the only and fully satisfying explanation. If one were to recognise this principle as the fundamental criterion upon which the editors of the individual codes directed themselves, it would follow to admit that those parts of the body which do not appear on a given list were consciously passed over as the damage to them was considered to be insignificant. However, there is no certainty as to whether such a conclusion is correct. For one cannot exclude that in concrete cases other factors did not influence the shaping of the said damaged bodily objects. For we can observe in the regulations also the appearance of parts of the body for which damage or even loss had a minimal influence on the life of the victim (for example, single teeth, hair, an eyebrow). Another significant criterion defining the choice of body parts as listed in the *leges* would be whether the given crime resulted in a visible change in the body's appearance.

The problem herein under consideration cannot be, so it seems, resolved only through the help of a hypothesis as to the role of a model catalogue for body parts or merely upon the basis of an assumption that the deciding factor was the damage to the body and its effects. We are rather dealing with a specific coupling of both of these elements within the process of creating a catalogue of crimes against the body. It follows here to add that in the case of each code the interaction of these factors has resulted in a different effect. In order to evaluate what role they played in the analysed process of bodily violation and damage, we need to elaborate on the various types of this crime.

Types of Crime against the Body

Any discussion of the issue of those crimes against the body described in the set of a dozen or so provisions from the laws of the Germanic peoples should begin with the assertion that we are dealing with a huge collection of individual cases that display many similarities. After all, the same or very similar kinds of injuries to the same body parts (e.g. finger removal, skull damage etc.) appear from code to code. What is more, combinations of crime types are similar. The recurrence of common features of this type is so evident that there would appear to be little point in discussing each law code separately. Furthermore, a description of the set of provisions contained in a selected code (e.g. in the Pact of Salic Law) may provide us with knowledge about the specific nature of this particular set of crimes.[256] Yet this does not contribute to our understanding of the character of the other texts or the features shared by the whole set. Hence the most effective method of researching the source material that is our focus of interest would appear to be the description of individual crime types appearing in the discussed set of Germanic laws.

Historians dealing with the *leges* have attempted for many years to draw attention to groups of crimes that are of the same or a similar nature. In the beginning, this was the preserve of legal historians. These studies have a long tradition stretching back to the 1840s. Wilhelm Wilda's work dating from this period, *Das Strafrecht der Germanen*, contains, I believe, the first academic description of various types of crime against the body in the Germanic *leges*.[257] He iso-

---

256 An interesting analysis of the complete array of provisions relating to crimes against the body was recently presented by Lisi Oliver in the *Æthelberht Laws* (*Beginnings...*, pp. 99–105). However, the research topic was not so much separate kinds of injury as their influence on the condition of particular body parts or organs, hence the state of health, mobility and perceptional capabilities of the injured party. This type of medical assessment of these injuries is compared with the level of compensation payments assigned to concrete cases.

257 W. E. Wilda, *Das Strafrecht der Germanen*, Halle 1842, p. 731; cf. L. Günther, *Ueber die Hauptstadien der geschichtlichen Entwicklung des Verbrechens der Körperverletzung und seiner Bestrafung*, Erlangen 1884, p. 47.

lated three main ways of causing bodily injury: wounds (*Wunden*), blows (*Schläge*) and causing paralysis (*Lähmungen*). He was alluding to the classification into these crimes which was included in the *Epilogue* of title XXII of the *Lex Frisionum*. There we read of *vulneribus* (wounds), *percussionibus* (blows) and *mancationibus* (injuries causing paresis or paralysis).[258] Wilda isolated eight sub-categories from the three primary categories, and these were sometimes divided into concrete crime types. These were: 'basic' injuries that could be measured (i.e., they were quantifiable), injuries requiring a doctor's special treatment, deep penetrative wounds (including injuries to the brain (or head), abdominal cavity or bone injuries), punctures that went right through the body, bone fractures (both closed and open), injuries causing disfigurement and scars, and finally those which caused paresis or physical disability.[259]

L. Günther, who reviewed Wilda's findings, added in his commentary to the above classification that the Germanic legal sources, despite their diversity, displayed many common features with regard to the crime types that appeared in them. Nevertheless, at the same time there are marked differences with regard to the levels of compensation stipulated in the case of each crime. He also draws attention to the wide variation in the values of individual compositions within any given code of tribal law serving as a criterion facilitating the definition of different categories of injury.[260] The distinctive features of the various types of bodily injury presented by the cited authors were not considered to be as important as research on crimes as a subject of study in the history of penal law.

Anette Neiderhellmann adopted different criteria as her point of departure when creating a classification system for acts of this nature in a work relating to the medical aspect of the provisions that concern us in the Germanic *leges*. The author lists four criteria which serve in the definition of physical injuries: type of injuring tool, type of injuring action, detailed location (e.g. an injury to the head) and clinical inventory (course of treatment or consequences of the injury).[261] She went on to describe nine primary categories of bodily injury: 1) general, 2) skin-related, 3) oedemas, 4) bloody injuries, 5) bone injuries, 6) skull damage, 7) injuries to the ribs and internal organs, 8) injuries causing impairment of organic function, and 9) injury after-effects. Each of these categories covered specific,

---

258 *LFris*, XXII, *Epilogus: Liti vero compositio, sive in vulneribus sive in percussionibus sive in mancationibus et omnibus superius descriptis* [...].
259 W. Wilda, *op cit*., pp. 734–774; cited after: L. Günther, *op. cit.*, pp. 50–57.
260 L. Günther, *op. cit.*, p. 57 ff.
261 A. Niederhellmann, *Arzt...*, p. 207.

concrete cases which were expressed in the texts of each code using separate Old Germanic terms (glosses) and described more broadly in Latin.[262]

A completely different approach to differentiating the crimes has recently been proposed by Lisi Oliver. In her analysis of the provisions regarding the various *personal injuries* appearing in the *Laws of Æthelberht*, she investigated their character from the perspective of two criteria, which were then used as a basis for determining the level of compensation payable upon a crime being committed – i.e., the value of a given part of the body from a physiological perspective and the extent to which the effects of a given injury were visible. She goes on to postulate two crime categories: those that mainly entail damage to the body substance (*injuries*) and those which may be viewed as *insults* manifesting themselves in alterations to the injured party's appearance. This manner of description, although it avoids 'technical details' when differentiating bodily injuries, should not be treated as a simplification.

The proposed classification of crimes against human bodily integrity were created, as we have maintained, from the point of view of research into the history of penal law (Wilda and Günther) or in connection with studies on the medical aspect of the *leges* provisions that are the focus of our interest (Niederhellmann). The attempt to classify types of injury to the body which we present below is of a different nature, for we consider descriptions of separate types of breach of bodily integrity or inviolability in terms of the manner and extent to which they contributed to the exposure of body parts.

---

262 A. Niederhellmann (*ibidem*, pp. 208–297) has described about 40 Old German (Frankish, Alamannian, Bavarian, Lombard, Anglo-Saxon) terms denoting various types of bodily injury: *dolg* (injury, scar), *cladolg* (nail injury), *palcprust* (skin wound), *pulislac/bunislegi/ pulslahi* (a blow causing swelling), *trunckslac* (a 'dry blow' i.e., an injury which does not draw blood), *durslegi* (a semantic equivalent to *pulislag*), *uuadfalt/uuidfalt* (a blow, a bruise, a blow that causes bruising), *plotruns* (a light blood wound, bleeding), *adargrati* (vein injury), *peinschrot* (bone damage), *kepolsceini* (skull exposure), *inanbina ambilicae* (exposed bone), *chicsciofrit, chesfrido, charfrido, cusfredum* (blow or injury to the head), *hrevovunt, hrevavunt/hreuauunti* (stomach injury), *gorovunt/goravunt* (bowel injury), *ferchvunt* (mortal injury), *freobleto* (injury or damage to the abdominal cavity), *gisifrit* (rib injury), *sicti/secti* (cut), *frasito* (amputated nose), *chamin* (injury causing physical impairment), *lichavina* (removal of an eye or facial disfigurement), *inchlaviuna* (facial disfigurement or trauma), *channichlśra/fune chleura* (injury to the ear or the whole cheek region), *chuldachina* (injury to the body or laming caused through injury), *alchaltea/alchacio* (causing muteness), *scardi/litscardi/ lidiscardi/orscardi* (partial damage or deformation of a bodily member), *wlitiwam* (facial injury), *spido* (deep scarring), *smelido* (reducing a bodily member) *glasaugi* (visual impairment or eye damage).

It must be made clear that accurate identification of the nature of the various acts presented in the *leges* provisions is difficult in some instances. This is not only due to the somewhat unclear manner in which some descriptions are formulated or their use of ambiguous words, but also the fact that in these provisions the terms which appear for acts may present their 'technical' nature (e.g., dismemberment). The crime descriptions thus describe the 'formal' features of the investigated injuries and harm inflicted on the body. This merely provides a basis for us to isolate the primary injury categories. These served as a point of departure for the creation of descriptions of individual acts which differed in terms of the violated body part type. The following crime terms offer an example of this kind of variance: knocking or gouging out an eye and knocking out a tooth. Both were described using the same verb: *excutere*.[263] Sometimes this variance resulted from a particular crime's effect. In our material we do in fact encounter acts described as blows (e.g., to the head), which could cause skin bruising or abrasions, but also deafness; in both cases, the word used to describe this act is *percutere*.[264] Cuttings relating to both the skin (e.g., on the face) and the arms or bones were described using the verb *incidere*.[265] The amputation of different body parts (e.g. ears, hands, digits) is another case altogether. This particular case is described using the word *castrare*, which in some parts of the *leges* texts denoted damage to the genitals causing impotence or sterility, while in other parts it would appear that it denoted their amputation.[266] The ambiguous nature of this word is indicated by the fact that in Salic and Alamannian law, two other verbs appear, besides the term *castrare*, which denoted excision – *abscidere* and

---

[263] E. g. *PLSal*, XXIX, 1. *Si quis [...] oculum eiecerit aut excusserit [...]*; 16. *Si quis dentem alienam excusserit [...]*; *ERot*, 48. *Si quis alii oculum excusserit [...]*; 51. *Si quis alii dentem excusserit [...]*.

[264] E. g., *PLSal*, XVII, 2. *Si quis alterum de sagitta toxicata percutere uoluerit [...]*; 8. *Si quis ingenuus ingenuum de fuste percusserit, ut sanguis non exeat [...]*; *LFris*, XXII, 1. *Si quis alium per iram in capite percuserit, ut eum surdum efficiat [...]*; 5. *Si eum percusserit, ut testa apparat [...]*.

[265] *LFris*, XXII, 11. *Si summam rugam frontis quis ictu transversam inciderit [...]*; 18. *Si maxillam inciderit [...]*; *Si costam transversam inciderit [...]*.

[266] *PLSal*, XXIX, 17. *Si quis hominem ingenuum castrauerit [...]*; *LSal*, XLVIII, 14. *Si quis ingenuos ingenuum castrauerit [...]*; *LRib*, VI. *Si quis ingenuus ingenuum castraverit [...]*; XXVIII (*De castratione servorum*). *Si autem eum castraverit [...]*; *LAl*, LVII, 59. *Si autem castraverit [...]*; cf. A. Niederhellemann, *Arzt...*, p. 142 ff., the author claims that usage in the *leges* of the mediaeval words *castrare* and *castratio* indicates a lack of distinction between castration understood as the removal of the sex glands or pharmacological sterilisation and penis damage leading to impotence. Loss of male sexual capability (*impotentia coeundi*) was identified with impotence (*impotentia generandi*). The concept of 'castration' thus denoted general inability to reproduce.

tollere.²⁶⁷ These examples urge us to conclude that it is advisable not to attach too much weight, when attempting to determine the nature of particular crimes, to the 'technical' expressions used to denote bodily injuries.

The discussed practices are of crucial importance to our research. The crime classification system which we want to present should therefore not only take into account the nature of an act, but also the specific nature of the injured body part and the effect this act caused. The catalogue of crimes given below was collated on the basis of all the investigated codes. It covers every type of damage and injury to the body. These are diverse in nature and not only include grave and irreversible mutilations leading to physical disability or the impairment of motor and sensory functions, but also injuries that cause disfigurement or lead to the victim becoming an object of mockery, various kinds of reversible violation of the body substance and breaches of bodily inviolability.

The set of provisions relating to crimes against the body that are contained in all thirteen *leges* cover a total of almost 500 items. This material can be divided into a dozen or so groups of regulations containing the main types of crime recurring in all the law books. Nevertheless, this set of provisions contains a significant number of different crime variants. The discussion that follows of the basic categories of these acts does not include all the different variants of injuries to or violations of the body which are recorded in the codes. Let us therefore proceed to a presentation of the individual crime types and the resultant injuries to different body parts. We will discuss the following cases in turn: dismemberments, castration, damage to the sense and speech organs, those causing paralysis, puncture wounds/perforations, exposure, bone fractures, disfigurements, cuts, knocking out teeth, bloody injuries, binding, wounds that do not cause blood flow and breaches of human bodily inviolability.

Dismemberments mainly related to the upper and lower limbs, hands, feet and fingers. Apart from this, we also encounter amputations of the nose, ear and tongue. Most of the investigated codes made separate mention of the amputation of every finger.²⁶⁸ Hands and feet were the body parts that were often encoun-

---

267 *PLSal*, XXIX, 18. *Si uero ad integrum tulerit* [*uirilia*] [...]; *LAl*, LVII, 58. *Si quis alium genitalia tota absciderit* [...]; apart from this, in different laws we find different descriptions of genital damage (the penis, the testicles, the scrotum) in which the following terms appear: *testiculum excutere, veretrum abscidere, viriculam transcapulare, foliculum testium traicere*.

268 *PLSal*, XXIX, 4. *Si quis policem de manu* [...] *excusserit* [...] *mallobergo alachtamo hoc est* [...], 6. *Si uero secundo digito, id est unde sagittatur sagitta excusserit, mallobergo alachtamo briorotero sunt* [...]; 9. [...] *Sequenti uero digito excusserit, mallobergo taphano* [...] *Quarto uero digito qui excussus fuerit, mallobergomelachano* [...] *Minimo digito qui excussus fuerit, mallobergo minecleno* [...]; *LSal*, XLVIII, 3. *Si police de ma-*

tered in an dismemberment context.[269] These provisions presumably relate to whole hands or feet. Many codes also discuss the case of toe amputations.[270]

---

nu capulauerit, mallobergo chramine [...]; 4. *Si quis secundum digitum, unde sagitta trahitur, excusserit, mallobergo brioro* [...]; 5. *Si medianom digitum, unde sagittatur, excusserit, mallobergo thaphano* [...]; 6. *Si quarto digito exusserit, mallobergo melagno* [...]; 7. *Si minimus digitus excussus fuerit, mallobergo menecleno* [...]; *ERot*, 63. *Si quis alii policem de manu excusserit* [...]; 64. *Si quis alii secundum digitum demanu excusserit* [...]; 65. *Si quis alii tertium digitum de manu, quod est medianus, excusserit* [...]; 66. *Si quis alii quartum digitum excusserit* [...]; 67. *Si quis quintum digitum excusserit* [...]; *LBaiw*, IV, 11. *Si quis alicuius pollicem absciderit* [...] *Et si proximum a pollice vel minimum absciderit* [...] *Illosmedianos duos digitos* [...]; *LFris*, XXII, 28. *Si quis pollicem absciderit* [...]; 29. *Si indicem absciderit* [...]; 30. *Si medium absciderit* [...]; 31. *Si annularem absciderit* [...]; 32. *Si minimum absciderit* [...]; *LThur*, 19. *Qui policem absciderit* [...]; 20. *Si indicem et inpudicum* [...]; 21. *Si medicum etminimum* [...]; cf. L. Oliver, *Body*..., pp. 143-148: writes about the functions and value of the fingers: 'The most important unite on the hand is the thumb – the digit which allows human to grasp and use a tool effectively. (...) A spear-fighter or sword-man who loses his little finger will be at considerable disadvantage in time of battle, as he will find it extremely difficult to control his weapon. A more precise linguistic interpretation of *sagittur* 'arrow finger', makes it obvious that the term relates to the arrow rather than a bow. The name refers to the first finger of the left hand, whose knuckle, aligned with that of the thumb, provides a level platform from which to launch the arrow shaft (...)'

269   *PLSal*, XXIX, 1. *Si quis alterum manum uel pedem debilitauerit* [...]; 11: *Si uero ipse pedes excussus fuerit, mallobergo chuldachina sichte*; *LSal*,XLVIII, 2. *Si uero ipsa* [sc. manus] *excusserit, mallobergo chramere* [...]; 9. *Si uero pedis percussus fuerit, mallobergo chuldachina* [...]; *ERot*, 62. *Si quis alii manum absciderit* [...]; 68. *Si quis alii pedem excusserit* [...]; *PLAl*, X, 13. *Et si manus tota excussa fuerit* [...]; XI, 1. *Si quis alteri pedem truncaverit* [...]; *LAl*, LVII, 66. *Si totum pedem absciderit* [...]; *LBaiw*, IV, 9. *Si quis libero* [...] *manum vel pedem tulerit* [...]; *LRib*, V, 4. *Si manum excusserit* [...]; 8. *Si quis ingenuus ingenuum pedem excusserit* [...]; *LFris*, XXII, 27: *Si manus in ipsa iunctura qua brachio adhaeret abscissa fuerit* [...]; *LSax*, XI. *Similiter de manibus, de pedibus* [abscisis]; *LThur*, XV. *Manus et pes abscisus* [...].

270   *ERot*, 69. *Si quis alii policem de pede excusserit* [...]; 70. *Si quis alli secundum digitum de pedse excusserit* [...], 71. *Si tertium digitum pedis excusserit* [...], 72. *Si quartum digitum excusserit* [...]; 73. *Si quintum digitum pedis excusserit* [...]; *PLAl*, XI, 4. *Si quis alteri articulum policem truncaverit* [...]; 8. *Si alis articulus truncatus fuerit* [...]; *LAl*, LVII, 64: *Si autem articulus prior abscissus fuerit* [...]; 65. *Illi alii articuli, si abscisi fuerint toti, unusquisque* [...]; *LFris*, XXII, 63. *Si pollicem pedis absciderit* [...] *si proximum digitum* [...] *si tertium* [...] *si quatrum* [...]. *Add. Sap.Wul.*, II, 12. *Pollex pedis* [...], 13. *Proximus digitus pollici* [...]; 14. *Secundus* [...] *tertius* [...] *quartus* [...]; *LSax*, XIII: [...] *Pollex pedis* [...] *Tres articuli medii* [...] *Mininus articulus* [...]. In the laws of the Alamanns, the term used for digits apart from the ordinary *digitus* was *articulus*, which denoted member, bone, joint or part; here, however, it clearly appears in the sense of toe.

Somewhat less commonly encountered in the Germanic *leges* were dismemberments of an arm or leg. *Castration* was a characteristic variety of the discussed injury type. Most of the analysed codes contained provisions relating to the amputation of genitals.[271] Crimes entailing the removal of the most prized body parts were valued as being equivalent to 50% in the case of a hand, foot or eye and 100%, in the case of a penis, of the *wergeld* for freemen stipulated in individual codes.[272] In the case of an ear or nose, the value of the composition was less (from 1/8 to 1/2 the value of the *wergeld*), while for fingers it was obviously less (for a thumb from 25 to 5%; for other fingers from 18 to 6%). These examples indicate that the most important determinant dictating the level of compensatory damages was the nature of the injured body part.

One group of crimes that is similar to amputations in terms of final outcome covers injuries causing the physical *impairment* or even *destruction of the organs of sensory perception or speech*. The specific type of injury entailing the victim being deprived of body parts was the gouging out of one or both eyes.[273] Loss of vision, according to some provisions, resulted from a blow to the eye.[274] Acts causing deafness or muteness also belong to this category. These could have been the outcome of an eye or tongue being cut out, but also a blow to the head.[275] The value of compensatory damages for causing deafness amounted to

---

271 Cf. A. Niederhellmann, *Arzt...*, pp. 142–154; L. Oliver, *Body...*, p. 128, writes: 'The genitals show more variation in the specifics of the damage: most common are rulings against damage to the penis (...) followed by testicles (...) scrotum (...) and loins. The low generally do not distinguish between castration and amputation of the penis: the crucial disability is the loss of the ability to create offspring.'

272 Usually, the level of these rates of compensation (in the case of a foot, hand or eye) was 50% of the *wergeld*; in the case of the amputation of a member, these payments were usually equivalent to 100% of the price of a man's head.

273 *PLSal*, XXIX, 12. *Si quis alteri oculum euellerit, mallobergo lichauina* [...]; *LSal* (D) XLVIII, 10. *Si quis alterum oculum eiecerit* [...]; (S) XVI, 13. *Si quis alteri oculum euellerit*; *LRib*, V, 3. *Si quis ingenuus ingenuum oculum excusserit* [...]; *ERot*, 48. *Si quis alii oculum excusserit* [...]; *PLAl*, V, 2. *Si oculus ille foras exierit* [...]; *LAl*, LVII, 7. *Si autem ipse visus foris exiit et ad melus* [...]; *LBaiw*, IV, 9. *Si quis libero oculum eruerit* [...]; *LThur*, 12. *Oculus unus vel ambo excusssi* [...]; *LSax*, 11. *Qui oculum unum excusserit* [...] *Si ambos* [...]; *LFris*, XII, 46. *Si totum oculum eruerit* [...].

274 *LRib*, V, 3. *Si visum in occulum restiterit, et videre non poterat* [...]; *LFris*, XXII, 45. *Si quis oculum alterius ita percusserit, ut eo ulterius videre non possit* [...].

275 *PLSal*, XXIX, 15. *Si quis linguam alteri capulauerit, ut loqui non possit, mallobergo alchaltea* [...]; *LSal*, (D) XLVIII, 12. *Si quis alterius linguam capulauerit, unde loquere non possit*; (S) XVI, 16. *Si quis linguam alterius amputauerit, ut loqui non possit* [...]; *LAl*, LVII, 9. *Si autem sic [aurem] absciderit profundo et eum exsurdaverit* [...]; 26. *Si autem lingua tota abscisa fuerit* [...] *Si autem media ut aliquid intellegat quod loquitur*

134   Chapter Three

from 25 to 100% and the damages for causing muteness amounted to from 25 to 50% of the sum adopted in individual codifications as the *wergeld* for free men.[276]

The crime which constituted grave bodily harm was *acts causing the paralysis* of particular body parts. Often the listed body parts that had become lame as a result of acts of violence were the hands and feet.[277] One particular variety of a crime of this kind was injury to the genitals causing impotence.[278] Provisions relating to the paralysis of fingers and toes are also quite often encountered. In most codes, compensation for causing paralysis amounted to half of the sum incurred for amputating the body part in question.[279]

The next crime category appearing in the *leges barbarorum* relates to *puncture wounds* to various parts of the body. This is a bodily injury that appears fre-

---

[...]; *LFris*, XXII, 1. *Si quis alterum per iram in capite percusserit, ut eum surdum efficiat* [...]; 2. *Si mutus efficiatur, sed tamen audire possit* [...].

276 Cf. L. Oliver, *Body*..., p. 93: 'The most usual pattern for compensation is that eye = nose = ear. This is the extent of regulation in Salic and Thuringian law, and I would be inclined to take such a ruling as implicit for Burgundian and Chamavan law also. Ripuarians, Alamanns, Bavarians, and Saxons all add to the equation a half fine either if the organs is struck off but the sensory perception remains, or if the sensory perception is damaged but the organ remains.'

277 *PLSal*, XXIX, 2. *Si cui uero manum capulauerit ut mancus pependerit, mallobergo chaminus hoc est* [...]; 10. *Si uero pedes capulatus fuerit et ibidemmancatus tenuerit, mallobergo chuldachina chamin* [...]; *LSal*, XLVIII, 1. *Si quis alterum manum capulauerit, unde homo mancus sit et ipsa manus super eum pendat, mallobergo secti* [...]; 8. *Si uero pedis capulatus fuerit et ibidem manaus teniat, mallobergo chudachina chamina* [...]; *ERot*, 62. [...] *et si [manum] sideraverit et non perexcusserit a corpore* [...]; *PLAl*, XI, 2. *Et si (pes) mancat* [...]; *LFris.*, XXII, 76. *Si manus percussa manca pependerit* [...] *pes similiter* [...]; *LSax*, XII. [...] *manus vel pes ita percussa manca loco remanserint* [...]; *LThur*, XV. *Manus vel pes* [...] *si manca pependerint* [...].

278 *PLSal*, XXIX, 17. *Si quis hominem ingenuum castruerit aut uiriculam suam transcapulauerit, unde mancus sit, mallobergo gaferit sunt* [...]; *LSal*, XLVIII (D), 14. *Si quis ingenuos ingenuum castrauerit aut uirilem transcapulauerit, unde mancu sit, mallobergo uuiderdi* [...]; cf. A. Niederhellmann, *op. cit.*, p. 149 ff., where the meaning of the mallberg glosses is explained – *gasferit* (injury to a member or castration) and *uuidardi* (injury to a man).

279 Under Bavarian law, a restitution payment of 40 *solidi* was prescribed for the amputation of a foot or hand and 20 *solidi* for causing their paralysis (the *wergeld* for a free man was 80 *solidi*). Under Thuringian law, compensation in analogous cases amounted to 300 and 150 *solidi* (the *wergeld* of an *adaling* (nobleman) – 600 solidi) In the *Lex Saxonum* – in the case of a foot – it was 720 solidi for amputation and 360 for causing paralysis (the *wergeld* for an adaling – 1440 solidi); in the *Lex Ribuaria*, amputation of a thumb entitled the victim to a restitution payment of 50 solidi, and causing paralysis of the same, 25 solidi (*wergeld* – 200 solidi).

quently and in many codes. It related to the head (or skull), trunk, limbs and internal organs. The law codes of individual tribes often mention puncture wounds to the arms and legs.[280] Sometimes additional attention was drawn to thigh and shin injuries[281], as well as those to the upper and lower elbow.[282] One feature that is common to all the descriptions of these injuries is that there is no mention in them of bone injury. This could appear surprising if one considers that they affected the arms and legs. This is particularly puzzling in the case of the elbow. The Germanic *leges* also quite often mention puncture wounds to the trunk.[283] The level of compensation in the puncture wounds group varied depending on the type of injured body part (trunk/innards, face, limbs).[284] The highest values (45, 28, 25%) were ascribed to puncture wounds or injuries to the innards (the abdominal cavity) as well as the nose and ear.

A separate group of crimes was made up by *bone fractures*. The investigated law codes mention injuries to bones in the skull, arms and legs.[285] Besides this, in several codes, rules were recorded relating to fractures to unspecified bones.[286] In the case of fractures to bones in the arms and legs, the level of compensation in relation to *wergelds* was quite low – from 5 to 22.5%. A specific

---

280 *LFris*, XXII, 84. *Si quis brachium vel coxam alterius transpunxerit* [...]; *LSax*, V. *Si* [...] *coxam vel brachium perforaverit* [...]; *ERot*, 57. *Si quis alium in brachio punxerit et transforaverit* [...]; 60. *Si quis alium in coxa plagaverit aut punxerit, si transforatum fuerit* [...]; *LAl*, LVII, 60. *Si quis alium ambas coxas uno iectu transpunxerit* [...].

281 *PLAl*, VII, 3. [...] *si coxa transpunctata fuerit* [...]; 4. *Si subtus ienuculum* [...]; *LAl*, LVII, 63. *Si tibia subtus genaculo transtuncta fuerit* [...]; *ERot*, 60. *Si quis alium in coxa plagaverit aut punxerit, si transforatum fuerit* [...].

282 *PLAl*, VII, 1. *Si quis brachium super cubito transpunxerit* [...]; 2. *Si subtus cubitum fuerit* [...]; *LAl*, LVII, 31. *Si quis alium brachium super cubitum transpunxerit* [...]; *LBaiw*, IV, 12. *Si quis alicui brachium supra cubitum transpunxerit* [...] *si ante cubitum transpunxerit* [...].

283 Cf. L. Oliver, *Body*..., p. 112, remarks: 'Curiously, regulation of wounds to the torso in barbarian law generally do not receive the detailed attention given to the head or the fingers. (...) Frisia, Alamannia, Wessex, and Kent – demonstrates considerably more interest in legislating injury to the torso than the rest. '

284 *Æthel*, 46. *Gif hit sio an hleore* [...]; 47. *Gif butu ðyrele sien* [...]; i.e., the puncturing of one or both cheeks; *LFris*, XXII, 47. *Si quis alium pectus forverit* [...]; 49. *Si praecordia perforaverit* [...]; 85. *Qui maxillas utrasque cum lingua sagitta vel quolibet telo transfixerit* [...]; *Add. Sap*, III, 31. *Si stomachus vel botellus perforatus fuerit* [...].

285 *PLAl*, IX, 1. *Si quis alteri brachium super cubitum aut coxa super ienuculum* [...] *frigerit* [...]; *LAL*, LVII, 35. *Si enim brachium fregerit, it aut pellem non rumpit* [...]; *LFris*, XXII, 24. *Si brachium ictu supra cubitum confractum fuerit* [...]; 25. *Si infra cubitum unum ossium confractum fuerit* [...]; 26. *Si utraque ossa fracta fuerit* [...]; *LFris*, XXII, 60. *Si coxam supra genu vulneraverit et os transverum fregerit* [...].

286 *LRib*, III. *Si quis ingenuus ingenuum in quolibet membro osso fregit* [...].

crime type was blows to the head leading to the shattering of the skull and exposure of the meninx. In the investigated material, several types of skull injury can be distinguished. Usually, they were fractures or punctures to the bone tissue. There are also several recurrences of injuries involving a skull fracture in which one or more bone splinters/fragments have exited from the wound.[287] Due to both the marked severity of this type of injury and the importance of the injured body part (fractures or punctures to the skull could cause exposure of the membrane covering the brain or even the emergence of the brain itself), the level of compensation (45, 33, 26% of the *wergeld*) was closer to that which appeared in the case of some amputations or injuries causing paralysis.

Another crime type is comprised of various types of *facial disfigurement*. These impairments, although they were injuries, also left permanent, irremovable traces on the body. In some cases, the investigated law codes do not go into detail about the effects of a given act on the appearance of the injured party's face. Here we have clauses giving information about injuries to the upper or lower lip, the upper or lower eyelid, the forehead, eyebrows[288] and puncture

---

287 L*Visig*, IV, 4, 1. *Si ingenuus ingenuum quolibet hictu in capite percusserit* [...] *pro osso fracto* [...]; PL*Sal*, XVII, 4. *Si quis alterum in caput ita plagauerit, ut cerebrum appareat* [...] *mallobergo chicsiofrit* [...]; 5. *Et si exinde tria ossa, qui super ipso cerebro iacent, exierint* [...]; E*Rot*, 47. *Si quis alium plagaverit in caput, ut ossa rumpantur, pro uno osso conponat solidos duodecim; si duo fuerint conponat solidos viginti et quattuor; si tres ossas fuerint conponat solidos tringenta et sex* [...]; PL*Al*, I, 4. *Si quis alteri caput frangit, sic ut ossus de capite ipssius tollatur* [...]; 5. *Si talis corpus fuerit, ut [de ca]pite ossus radatur et frangit* [...]; L*Al*, LVII, 4. *Si autem de capite ossum fractum de plaga tullerit* [...]; 6. *Si autem testa trescapulata fuerit, ita ut cervella appareant* [...]; 7. *Si autem ex ipsa plaga cervella exierunt* [...]; L*Baiw*, IV, 5. *Si os tulerit de plaga de capite* [...]; 6. *Si cervella in capite appareant* [...]; L*Fris*, XXII, 6. *Si os [testae] perforatum fuerit* [...]; 7. *Si membranam, quam cerebrum continetur, gladius tetigerit* [...]; cf. L. Oliver, *Body*..., pp. 75 and 86-87: 'As a rule, there is a distinct division between the insular laws – regulating damage only to the tabula – and the continental laws – distinguishing between spall and exposure of the brain. Frisia adds a third category: the laws differentiate between touching the 'membrane' (probably the *dura mater* – the tough, rubbery membrane immediately inside the skull – implying a break of both outer and inner tabula) and exposing the brain (which must then logically mean piercing the *dura* and probably *pia mater* – the finer membrane that directly surrounds the brain. (...) Frisian law recognizes the difference between the two bony layers of the skull, as is the case in Anglo-Saxon England, but extends legislative rulings to the scalp and the spilling of blood, evidenced only in continental law.'

288 L*Al*, LVII, 11. *Si enim superior palpebris maculata fuerit, ut claudere non possit* [...]; 12. *Si enim subterior maculata fuerit, ut lacrime continere non possit* [...]; 18. *Si enim labium superiorem alicuis maculaverit, ita ut dentes appareant* [...]; 19. *Et subteriorem, ut salivam continere non possit* [...]; L*Baiw*, IV, 15. *Labia subteriore* [...] *et palpebre*

wounds or other damage to the nose[289], ear[290] or cheek.[291] The effects of the mentioned acts can only be described through the interpretation of individual clauses. The restitution percentage in relation to *wergelds* varied from code to code in the case of facial disfigurement, but was generally quite low (from 3.75 to 25%).[292] In the investigated provisions, we often encounter cases of teeth being knocked out. In particular codes, variants of this crime have been mentioned that differ in terms of the type of affected tooth (front, upper, lower, side, molars).[293] In some codes, cases of (visible) front teeth being knocked out causing

---

     subteriore [...] *Superiorem vero palpebram vel superiorem labiam si maculaverit* [...]; *LFris*, XXII, 11. *Si summam rugam frontis quis ictu transversam inciderit* [...] oraz 12–13, 14. *Si supercilium inciderit* [...]; 15. *Si palpebram, aut superiorem aut subteriorem, vulneraverit* [...].

289 *Æthel*, 45. *Gif nasu ðyrel weorð* [...]; *LBaiw*, IV, 13. *Si quis nasum transpunxerit* [...]; *LFris*, XXII, 16. *Si nasum transpunxerit* [...]; *Add. Sap. Wul*, II, 21. *Si nasus una parte perforatus fuerit* [...]; 22. *Si cartilago perforata fuerit* [...]; 23. *Si etiam ex altera parte telum exierit, ita ut III foramina facta sint* [...].

290 *Æthel*, 45. *Gif nasu ðyrel weorð* [...]; *LBaiw*, IV, 13. *Si quis nasum transpunxerit* [...]; *LFris*, XXII, 16. *Si nasum transpunxerit* [...]; *Add. Sap.Wul*, II, 21. *Si nasus una parte perforatus fuerit* [...]; 22. *Si cartilago perforata fuerit* [...]; 23. *Si etiam ex altera parte telum exierit, ita ut III foramina facta sint* [...].

291 *Æthel*, 46. *Gif hit sio an hleore* [...]; *LFris*, XXII, 18. *Si maxillam inciderit* [...].

292 Cf. L. Oliver, *Body*..., p. 101, writes about injuries of facial features: 'The North Sea grouping [i. e. Anglo-Saxon and Frisian laws – my addition] (...) makes no mention of functionality. (...) Clearly these fines primarily consider cosmetic rather than physiological harm. (...) The only other early medieval laws to give such prominence to visible damage to the face – that is, insult added to injury – are those of the Irish, who regulate injury to eyelash, eyebrow, cheek, chin, and jawbone. The evidence from legal texts suggests that early medieval Frisia and Ireland place a higher value on insulting visible but only superficially harmful damage did their contemporaneous neighbours, who legislated primarily for functional damage.

293 *PLSal*, XXIX, 16. *Si quis dentem alienam excusserit, mallobergo inchlauina* [...]; *Æthel*, 51. *Æt þamfeower toðum fyrestatum, æt gehwylcum VI scillingas: se toþ se þanne bistaneþ IIII scill'; Se ðe ðonne bi ðam standeþ III scilll'; ond þonne siþþan gewylce scilling*; *LAl*, LVII, 20. *Si enim aliquis alium uno hictu duos dentes superiores primas excusserit* [...]; 22. *Si autem dentem absciderit, quod „marczan" dicunt Alamanni* [...]; 23. *De alias vero de qualecumque excusserit* [...]; 24. *De superioribus vero duabus primas alicui exusserit* [...]; *LBaiw*, IV, 16. *Si quis alicui dentem maxillarem, quod marchzand vocant, excusserit* [...]; *Alios vero dentes si excusserit, ununquemque* [...]; *LFris*, XXII, 19. *Si unum dentem de anterioribus excusssserit* [...]; 20. *Si unum de angularibus dentibus excusserit* [...]; 21. *Si de molaribus unum excusserit* [...]: Cf. L. Oliver, *Body*..., p. 102: 'In terms of physiological function, the most valuable teeth in the mouth are the canines. (...) The second-most utile teeth are the molars, which grind and

disfigurement are treated separately. However, in most codifications this crime was classified in the same category.

Another group of injuries were *cuts* to various body parts. These acts differed from each other in terms of the degree of interference to the body substance. Some of them were 'surface' in nature, while others were deep. Those that had relatively less severe repercussions for the victim would appear to include cuts to the neck or shoulder (*iugulum*)[294] and the upper and middle finger joints.[295] These were mainly skin injuries. The most serious cases involving cuts encroached on the bones, muscles and internal organs.

In many law codes, *bloody injuries* are classified as a separate crime type. The term injury with the effusion of blood (*sanguinis effusio*)[296] is often used to describe them; sometimes there is also mention of blood pouring out onto the ground.[297] The appearance of blood as the result of an injury or blow was crucially important here as a criterion for distinguishing this crime. In definitions of this kind of crime the fact of blood pouring out or bleeding was accentuated. In the *Lex Baiwariorum*, there is mention of severed veins and the bleeding caused by this.[298]

Another act mentioned in the *leges barbarorum* which constituted *a breach of personal inviolability* was tying a person up. Slapping a man in the face (*alapa*) constituted a breach of personal inviolability and at the same time a direct attack on the body substance. The group of crimes categorised as violations to the body substance comprised various types of battery which failed to cause wounding, but left their trace in the form of bruises or swelling. One sporadically encountered kind of bodily violation was injury causing the appearance of undefined bodily fluid (*humor*).[299]

---

process food. (...) The least necessary teeth are the incisors, whose role is to tear or rip food – a task which can easily be performed by a *seax* or short knife..'

294 *LFris*, XXII, 22. *Si iugulum incisum fuerit* [...].
295 *Add. Sap.*, II, 7. *Si digitus quilibet superiori articulo praecisus fuerit* [...]; 8. *Si in subteriori praecisus fuerit* [...].
296 *PLSal*, XVII, 9. *Si uero sanguis exierit* [...]; *LSal*, XXIII, 3. *Si vero sangius exierit* [...]; *LBaiw*, IV, 2. *Si in eum sanguinem fuderit* [...]; *LRib*, XX, 1. *Si servus ingenuo sanguine effusione fecerit aut regie vel ecclesiastico homine* [...]; *LThur*, V. *Sanguinis effusio* [...]; *LSax*, 3. *Si sanguinat* [...]; *LFris*, XXII, 4. *Si autem sanguinem fuderit* [...].
297 *PLSal*, XVII, 3. *Si quius hominem plagaverit ita, ut sanguis ad terra cadat* [...]; *LRib*, 11. *Si quis ingenuus ingenuum percusserit, ut sanguis exiat, terra tangat* [...]; *LAl*, LVII, 2. *Si autem sanguinem federit, ut terra tangat* [...].
298 *LBaiw*, IV, 4. *Si in eo vena percusserit, ut sine igne sanguinem stagnare non possit* [...].
299 *LFris*, XXII, 35. *Si quislibet digitus ex quatuor longioribus in superioris articuli iunctura ita percussus fuerit, ut humor ex vulnere decurrat* [...]; similarly, clause 64 on toe injuries; cf. A. Niederhellmann, *Arzt...*, pp. 196–200 and 203–206.

Another type of bodily violation was the *cutting of hair or a beard*.[300] A separate crime type was comprised of acts involving the touching of the victim's body, assuming a lesser or greater use of physical force. However, in the majority of cases these acts caused temporary violations rather than damage to the body substance. Another crime against the body was, in the light of some law codes, the pulling (or grabbing) of a free man by his hair.[301] Such an act did not constitute a direct breach of bodily integrity. It did not cause permanent or even temporary damage, understood in the strict sense, to the body substance. The crime or harm that was caused in this case would appear to entail the breaching of the boundaries of a free man's bodily inviolability.

It should be stressed that the catalogue of bodily injuries above does not appear in such a developed form in any of the codes here investigated. Every code of barbarian laws possessed its own set of crimes of this type. However, the differences between them were relatively modest, and certainly significantly smaller than in the case of body parts. This is borne out by a comparison of data from the *Lex Ribuaria* and *Lex Frisionum*, i.e., the codes with a radically different number of provisions relating to crimes against the body (15 and 87, respectively). The similarities were not only of a qualitative nature, for they are obvious in relation to crime types.

Frisians (12):

dismemberments/amputations (16), injuries (16), cuts (10), puncture wounds (9), fractures (4), blows (3), knockings out (2), injuries causing incapacity/paralysis (1), grabbings (1), injuries causing deafness or muteness, tyings up (1).

Ripuarian Franks (8):

dismemberments/amputations (6), injuries causing deafness/incapacity (3), fractures (1), puncture wounds (1), injuries (1), blows (3), knockings out (1).

Alamanns (Lex) (9):

amputations (20), knockings out (5), puncture wounds (8), cases of incapacity/paralysis (2), fractures (3), disfigurations (4), shaving, cutting through (1).

These data contrasted with the number of body parts in the individual *leges* also indicate that we are not dealing in the investigated set of laws with an interrelationship of a kind in which a rise in the number of crimes causes a correspond-

---

300 *LAl*, LVII, 29. *Si quis alium contra legem tunderit caput liberum non volentem* [...]; 30. *Si enim barba alicuius non volentem tunderit* [...]; *LFris*, XXII, 17. *Si granonem ictu percusam praeciderit* [...].

301 *LFris*, XXII, 65 – grabbing by the hair in anger; *LSax.*, 7 – grabbing by the hair; *LConst*, V, 4 – cases of grabbing using one hand and two hands were treated separately; this provision also applied to freedmen and slaves.

140   Chapter Three

ing rise in the number of body parts. For in the mentioned codes no correlation of this kind appears between both sets.

* * *

The catalogue of bodily injuries presented above was reconstructed on the basis of provisions which primarily applied to freemen. In some of the cited law codes there were provisions that stated that the lists of crimes against the body that they contained were also meant to be applied in relation to women.[302] It therefore appears that the creators of these codes treated a woman's body as a kind of imitation of a man's. This was not, however, a common practice. In some *leges* (the Salian, Alamannian and Frisian laws), alongside regulations created with the male population in mind, there are also provisions which exclusively pertain to acts directed against the bodily inviolability or integrity of women.

These were mainly acts which entailed touching the victim's body, assuming a lesser or greater use of physical force. These acts in most cases, however, caused temporary violations rather than damage to the body substance. In the *Lex Frisionum* cases are mentioned of the forbidden squeezing of a free woman's breasts or the grasping of her vulva.[303] In both cases, the stipulation was added that these provisions applied to a woman who did not belong to the family *(femina [...] non sua)*. It should be added that in the Frisian law the cited provisions were two final points appended to title XXII, which dealt with several dozen crimes against various parts of a man's body.

Somewhat similar regulations, but located in a completely separate title, can also be found in Salian law *(Pactus* and *Lex)*. There is mention there of touching a woman's fingers, hand, arm and breasts.[304] In title XX of the *Pactus legis Salicae*, we first read of squeezing *(strinxerit)* the hand, shoulder or fingers. In the following paragraph, the verb *premere (brachium presserit)*, which means 'embrace' or 'squeeze' was employed in relation to the forearm. By contrast, the word *mittere* was used in relation to the arm above the elbow. It appears that the

---

302   This was the case in the laws of the Visigoths *(LVisig,* VI, 4, 3; p. 266), Anglo-Saxons (unmarried women), *Æthel,* 74; Alamanns *(LAl,* LIX, 2), Bavarians *(LBaiw,* IV, 30) and Saxons (girls), *LSax,* XV.

303   *LFris,* XXII, 88. *Si quis liberam feminam, et non suam, per mamillam strinxerit* [...]; 89. *Si per verenda eius comprehenderit* [...].

304   *PLSal,* XX, 1. *Si quis ingenuus homo ingenua mulieris vel qualibet feminae manum vel brachium aut digitum strinxerit* [...] *(chamin);* 2. *Si quis brachium presserit* [...] *(chamin);* 3. *Certe Si super cubitum manu miserit* [...] *(chamin malicharde);* 4. *Si quis mamillam mulierem strinxerit, aut sciderit, quod sanguis egressus fuerit (de bructe)...* (manuscript C 5) 4. *Si quis mulierem mamellam capulaverit; LSal,* XXV, 1. *Si quis ingenuus ad feminam ingenuam digitum aut manum extrinxerit* [...]; 2. *Si brachium strinxerit* [...]; 3. *Si super cubitum eam strinxerit* [...]; 4. *Si mamillam exstrinxerit* [...].

term *super cubitum manum miserit* should be understood in the sense of 'laying a hand [on the arm] above the elbow'. However, in paragraph 4, which pertains to the breast, apart from the word *stringere*, the following descriptive formula appears: *sciderit, quod sanguis egressus fuerit*. The verb *scindere* means 'to rip', 'to tear', 'to break'. It would therefore appear that it described two acts here: squeezing the breasts and an injury causing the appearance of blood. In another manuscript of the *Pactus*, the word *capulare*, meaning 'to sever', 'to cut', 'to cleave'[305], was used in the paragraph pertaining to the breasts, which would complement the meaning of the word *scindere*. The aforementioned acts would appear to belong to various categories: the first was an act breaching bodily inviolability, while the second, a type of injury, violated the body substance. The manner in which paragraph 4 is formulated, would appear to attest to the fact that the described acts were treated as being akin to each other in the sense that in practice touching of a sexual nature (groping) could easily turn into a brutal assault.

Crimes against a woman's body also include acts which breached bodily inviolability rather than being harmful to the actual body substance. What is of concern here is illicit exposure, in the understanding of the investigated custom law, of a woman's body. A description of this kind of act was set down in title LVI of the *Lex Alamannorum*. Details are provided of the following acts: exposure of the head by force, exposure of the legs up to the knees, and also exposure of the genitals or buttocks.[306] It is quite typical that a link should be made in a one act title between exposing the head, or more precisely the hair and raising clothing to expose the lower part of the body. It would appear that the aforementioned acts were treated as violating the norms defining the situations in which a woman could be naked. It can be seen from the above provision that the definition of a situation in which a woman was naked or had been disrobed also covered the forced exposure of her hair. The manner in which the investigated title was compiled indicates that a woman's hair was perceived as one of her sexual attributes. The crimes described were clearly akin to sexual assault. This is confirmed by the fact that the last paragraph of the cited title speaks of the rape of a disrobed woman. This is further reinforced by the fact that the law creators not only accentuate touching of the body, but also its public exposure, as highlighted

---

305 Du Cange, *Glossarium mediae et infimae latinitatis*, Graz 1954, vol. III, p. 149, synonyms: *incidere, scindere, secare*. The examples given there, which come from different Germanic law codes, apply to: hair cutting (*crines capulare*), tree cutting (*arborem capulare*) and amputating feet (*pedem capulare*).

306 *LAl*, LVI, 1. *Si quis libera femina vadit itinere suo inter villas et obviavit eam aliquis per raptum denudat caput* [...] *Et si eius vestimenta levaverit, usque ad genucla denudat* [...] *Et si eam denudaverit, usque genitalia eius appareat vel posteriora* [...] *Si autem ad ea fornicaverit contra voluntate eius* [...].

in the following sentence: 'If he strips her to such an extent that her genitals will be visible' (*si eam denudaverit usque genitalia eius appareat*).

The discussed breaches of female bodily inviolability are crucially important to our deliberations, for they show that in the cited barbarian codifications the human body, or more precisely body parts, were perceived and defined within the context of the victim's sex. Whereas in the case of provisions that speak of the application, to women, of regulations pertaining to the violation of and injuries to the male body, we are rather dealing with the omission of differences between the sexes (e.g., in the *Lex Alamannorum*, there was an entry about the amputation of male genitals, which quite obviously does not apply to women)[307], in the quoted examples of breaches of the bodily inviolability of women, the sexual nature of the victim's body was highlighted. In the context of the crimes against women discussed here, the provisions speaking of castration or the causing of impotence may not only be interpreted as one of the types of breach of a man's body substance, but also as an offence against his sexuality. These reflections seem to lead us to the more general assertion that: the human body (whether it belonged to a man or woman), once an object of crime, was not only viewed by the creators of these laws as a set of anatomical components and motor/sensory functions, i.e., as an organism, but also within a gender context which went far beyond human bodily or biological attributes. The exposure or touching of body parts treated as the female gender attributes were, first and foremost acts, directed against her dignity and inviolability and entailed breaching the customary norms regulating the domain of relations between the sexes.

Descriptions of Crimes against Bodily Integrality as an Account on the Human Body

Having learnt the main types of crimes against the body and their particular variants we are able to return to the question of the influence of the said damage on the selection of the body parts enumerated in the regulations under investigation. One may state most generally that certain types of crime were connected with certain bodily objects. Breakage concerned, obviously, various types of bone, severing – the limbs or their components as well as other protruding fragments of the body, a somewhat similar scope is covered by acts resulting in paralysis,

---

307 *LAl*, LIX, 2. *Haec omnia compositio, quod ad viris iudicavimus, ad feminis eorum omnia dupliciter conponatur*. This rule refered to the regulations contained in title LVII, 1–69, LVIII and LIX, 2; here equally to point 58 of title LVII, which talks of damage to the male genitals; Cf. also *LBaiw*, IV, 30. *De feminis vero eorum, si aliquid de istis actis continxerit, omnia dupliciter conponuntur* [...] and IV, 16. *Si quis aliquem castraverit* [...].

striking, wounding while cutting referred most often to the skin. Upon this basis one may surmise that the recording within a given code of certain types of damage resulted in the inclusion in the said of a definite group of bodily objects.

However, it follows to clearly state that in the overwhelming majority of cases the organisational structure of statements concerning crimes was the set of body parts and not a catalogue of damage and violations to the body itself. In other words the individual types of crimes were ascribed to the list of body parts and not the other way round. The order of enumerating the said acts was closely connected with the vision of the human body adopted by the creators of the laws. As we have already stated earlier they understood the body in the meaning of a figure or profile of man in the literal understanding of the word. Such a view of the human body defined the order in the description of crimes against it. Nothing indicates also that the appearance of some type of crime automatically resulted in the consideration of a given body object.

Research into the body as the object of criminality should concern not only the individual components of crime but also their arrangement and the forms of information transfer used in them. In connection with this the following questions come to the fore: in what way do statements on various types of crime influence the utterance on the body itself? How did a given crime (damage, violation) disclose particular body parts? Beginning from general statements one may propose that the types of damage here outlined constituted kinds of 'frames' or frameworks defining the scope of the information on the subject of a body fragment appearing in a given regulation. This transfer was limited most frequently to a description of the degree to which a given act violated the body's substance. At the same time the type of crime defined the point or place on the body where this damage occurred.

Germanic law makers listed on the whole the direct effects of acts of violence in the form of injuries, severing etc. In the majority of regulations the type of violation was defined (e.g. the cutting off of a finger or the piercing of an arm), treating this type of information as a sufficient description of the nature of the act and its significance. However, there exists a small group of regulations in which, besides the part of the body that underwent damage, there appears in the description of the crime other bodily objects adjacent to it or separating it from other segments (e.g., joints) or in some other way connected with it.[308] The regulation on the striking-wounding of a place not specified more exactly contained the account of the flowing of blood or about a vein (artery?); the one about the piercing of the skull talks of the cerebral leptomeninx or about the brain itself;

---

308 On the striking a joint: *LFris*, XXII, 38. *Si in iunctura manus et brachii hoc evenerit*; 40. *Si in summitate qua brachium scapulae iungitur* [...]; 43. *Si ad iuncturam brachii et policis* [...].

the regulation on the wounding of the face talks about the part not covered in stubble or about the hair and beard; the regulation on the cutting of the lower of the lines on the forehead says that they are found 'next to the eye'.[309] The regulations on damage to the leg are divided into the part above and below the knee, while the arms above and below the elbow, in the case of the fingers there is talk of the joints.[310] However, here the knee, elbow and finger joints are not always themselves mentioned in these regulations as the object of crime.

On the other hand in certain *leges* regulations there is visible an opposite tendency. The descriptions of crimes contained the designation of one bodily object, while passing over the information that this act caused damage to one or several others. In certain codes there appears a series of regulations that concern the following damage to the body: 1) the cutting through of the skin on the head so as to expose the skull bone, 2) the cutting through of the skull so as to expose the brain, 3) the brain coming out through the wound.[311] In the second point there is no longer talk of the skin on the head being cut, while in the third that the skull bone was pierced. Another example of such a method of creating a description of a crime are the examples on damage to the innards. The matter here concerns wounds, for example to the stomach or intestines but without information about other damage (e.g., the piercing of the skin, damage to the abdominal cavity) which out of necessity came into being as a result of crimes of this sort. The individuals editing these regulations were most clearly sure that

---

309 *LBaiw*, IV, 4. *Si in eo venam percusserit* [...]; *PLSal*, XVII, 4. *Si quis alterum ita in caput ita plagauerit ut cerebrum appareat* [...] *mallobergo chicsiofrit* [...]; 5. *Et si exinde tria ossa, quis super cerebro iacent exierint, mallobergo chi(c)ssiofrit* [...]; *LSal*, XV(S), 4. *Si quis hominem in caput ita plagauerit, ut deinde tria ossa exeant* [...]; 5. *Si quis hominem ita plagauerit in caput, ut cerebrum appareat et tria ossa desuper cerebrum exierint* [...]; *ERot*, 46. *Si quis alii plagam in caput fecerit, ut cutica tantum rumpatur, quod capilli cooperiunt* [...]; 47. *Si quis alium plagaverit in caput, ut ossa rumpantur* [...]; *LAl*, LVII, 6. *Si autem testa trescapulata fuerit, ita ut cervella appareat* [...]; *LFris*, XXII, 7: *Si membranam, qua cerebrum continetur gladius tetigerit* [...]; 8. *Si ipsa membrana rupta fuerit, ita ut cerebrum exire possit* [...]; *LAl*, LVII (B), 27. *Si autem aliqua plaga in facie alicuis facta fuerit, quam capilli vel barba non cooperiant* [...]; *LFris*, XXII, 13. *Si tertiam* [rugam frontis], *quae iuxte oculos est* [...].
310 E.g., *LAl*, LVII, 35. *Si enim brachium fregerit* [...] *ante cubitum* [...]; 36. *Si autem super cubitum hoc contingerit* [...]; 37. *Si enim in cubitu percusus fuerit* [...]; 39. *Si autem a cubito absciderit*; 46. *Si longissimus digitus in primo nodo abscisus fuerit* [...]; 47. *Si in secundo nodo* [...]; *LFris*, XXII, 24. *Si brachium ictu supra cubitum confractum fuerit* [...]; 25. *Si infra cubitum unum ossium confractum fuerit*; 60. *Si coxam super genu vulneraverit* [...]; 61. *Si tibia subtus genuculo media inciasa fuerit* [...].
311 *LAl*, LVII, 3–4, 6–7; *LBaiw*, IV, 4–6; *LFris*, XXII, 5–8.

the potential recipient of the second regulation would understand it within the context of the first, and the third rule in the context of the first and second.

Depending on the type of crime and on the type of body part those editing the individual legal codes either disclosed the bodily 'context' of the given damage or passed over its existence. The first of these practices meant that in these regulations there occurred a definite small number of body parts constituting as if an additional account of the crime victim's body. This type of editorial action was not, however, frequent. There dominate descriptions where the subject is the only selected object, appearing as if independently, without the appearance of the accompanying 'background'.

* * *

A man's body, and specifically its individual parts were, as we have stated, perceived and presented in the regulations of the *leges* as if in the reflection of the crimes committed against it. Our research should evaluate not only the detailed content of the regulations but also the form of the utterances as used in them. When we look at the sets of regulations as a specific whole, we see that the said damages and violations were subordinated by those writing down the laws to the list of body parts contained in each of the codes. Sometimes as a result of this there arose a series of regulations that concerned various forms of damage to the very same part of the body (e.g., the skull, arms).

We shall refer to concrete examples in order to illustrate the importance of this phenomenon. In *Lex Alamannorum* we find 5 different, consecutively arranged points (LVII, 3–7) on types of damage to the head (wounds to the skin exposing the skull, the splitering of fragments of the skull bone, the piercing of the skull with exposure of the brain, the protruding of the brain from the broken skull)[312] as well as 10 regulations appearing one after another (LVII, 31–40) talking of types of damage to the arm (perforating above and below the elbow, the piercing of the hand so that blood was drawn, the closed facture of the bones of the arm above and below the elbow, striking the elbow causing paralysis, paralysis of the whole arm, severing of the arm from the elbow and from the shoulder).[313] A similar phenomenon appears in *Lex Frisionum*, where in four subsequent clauses (XXII, 5–8) are contained regulations on types of damage to the skull (i.e., exposure of the scalp, piercing of the skull, piercing the skull to the cerebral leptomeninx, piercing the cerebral leptomeninx); besides, in points 1 and 2 of this same title, there is mention of blows to the head – resulting in deafness and muteness. In the Frisian code we can find also two groups of regu-

---

312   *LAl*, LVII, 3–7.
313   *LAl*, LVII, 31–40.

lations on arms; the first (XXII, 24–27) covers four successive points (the breaking of arms above and below the elbow, the breaking of both parts of the arm, the breaking of the wrist), while the second (XXII, 38–43) – 6 regulations (wounding with the production of fluid from the wound; the wrist, elbow, shoulder joint, the upper thumb joint, the lower thumb joint, the joint connecting the thumb to the hand). In the *Laws of Æthelberht* (39–42) – 4 subsequent provisions concern the ear (the causing of deafness, cutting off, piercing, the cutting or partial cutting off of the ear).

The cited examples, drawn from the codifications containing the largest number of provisions of interest to us, show that we are herein dealing with a certain kind of block covering one fragment of the body and several different variants in the types of damage ascribed to it. In a significant number of the researched codes the said catalogues were arranged in an order from head to foot. The role of the set of body parts as an organising factor in accounts of crimes committed was important and decisive. In the *leges* in which a different way of enumerating the parts of the body was adopted their significance was, however, no less. In none of the *leges* under consideration did the catalogue of crimes against the body constitute an independent structure, intentionally utilised as a cognitive model organising the set of provisions of interest to us.

These observations are of real significance for the problem of the body as an object of crime. For they show it was the very described sets of body parts that constituted the first object of interest for those writting down the said codes. They defined in a sense their way of thinking, their perception of crimes which had taken in the researched *leges* a somewhat secondary place. This is a form of paradox, for the main subject of legal regulation is, as it seems, activity of a criminal nature, and not the object upon which this activity manifested itself.

The Effects of Crimes against the Body

Various descriptions of damage to the frame of the body although containing an account of its individual fragments are not the only textual mechanism which activated discourse on various aspects of a man's corporeal shell. Another element within the provisions under discussion are the pieces of information concerning various types of outcome resulting from particular crimes.

Reading the regulations on bodily crimes we quickly notice that in the overwhelming majority of cases the content is restricted to a statement of simple facts on the nature of the crimes committed. There is talk of blows, woundings, piercings and puncturings as well as the cutting off and severing of particular parts of the body. The majority of provisions contained in the Germanic *leges* concerned crimes involving harm inflicted on the body in the strict sense of the word. Generally the provisions analysed give no information about the consequences brought about by these acts of violence (e.g., the cutting off of a hand) for the victim themselves. Those responsible for the writing down of the provisions presumably took the view that these consequences are, in themselves, obvious; there being, consequently, no need to stipulate them each and every time. Some, few in number, codes of the Germanic laws do, however, contain information on the varied results of bodily violation experienced by the victims. These were presented in the form of laconic commentaries to the description of the given crime (e.g., if someone damaged someone else's arm in such a way that he/she could do nothing with it). These consequences could be divided into several categories as a result of their nature. These are: deformation of the body's appearance (particularly the face), the loss of the organism's higher functions linked to sensual perception and communication, the loss of physical ability and fitness.

We shall start by discussing the regulations for crimes that resulted in changes to the victim's face. *Lex Alamannorum* in particular contains many such regulations; this talks of the disfiguring of the upper eyelid in such a way that the eye cannot be closed, the lower eyelid such that it cannot stop watering, of damage to the upper lip so that the teeth are exposed as well as of the lower lip so that it is unable to retain saliva, equally the cutting off of the tip of the nose so that nasal secretions cannot be retained.[314] Similar provisions may be found in *Lex Baiwariorum*; here the matter concerns the disfiguring of the lower lip and the lower eyelid with the self same results that have been outlined above. Besides, this code contained also titles referring to disfiguration of the upper and lower lip and the ear, *ut exinde turpis appareat*.[315] In *Lex Ribuaria* we have found clauses concerning damage to the nose resulting in the victim's inability

---

314 *LAl*, LVII, 11. *Si enim superior palpebris maculata fuerit, ut claudere non possit* [...]; 12. *Si enim subteriore maculata fuerit ut lacrime continere non possit* [...]; 16. *Si enim summitatem nasi, ut muccus continere non possit* [...]; 18. *Si enim labium superiorem alicuius maculaverit, ita ut dentes appareant* [...]; 19. *Et subteriorem, ut salivam continere non possit* [...]; cf. L. Oliver, *Body*..., pp. 99-100: 'The two eastern-most territories [i. e. Bavaria and Alamannia] differentiate between the two eyelids on the basis of their functions: destroying the lower so that eye cannot contain tears draws a fine twice that for destroying the upper so that eye cannot close. These very specifically located injuries may well have been inflicted for the purposes of deliberate mutilation (...) if the eye cannot be covered, it remains unprotected from flying dust and debris. (No person suffering such an injury could ever, for example, ride a horse.)'.

315 *LBaiw*, IV, 15. [...] *Si sic plagaverit* [palpebrem], *ut lacrimam continere non possint, vel subterior labia salivam non continet* [...] *Superiorem vero palpebrem vel superiorem labiam. Si maculaverit* [...]; 14. [...] *Si aurem maculaverit, ut exinde turpis appareat, quod lidiscrati vocant* [...].

to blow it.³¹⁶ In turn, in the Lombard *Edictum Rotharii* there is contained a provision on the cutting off of the lips, which resulted in one, two or three teeth being made visible as well as a second clause wherein there is talk of the knocking out of one, two or more teeth, whose loss is visible when the victim smiled.³¹⁷

The facial damage outlined, besides the fact that it concerned a greater or lesser loss of body tissue itself, also resulted in physiological difficulties, such as: dribbling, watering, phlegm secretion, the inability to shut the eye. However, it seems that the most difficult problem was disfigurement of the face as well as the humiliating or shameful nature of these disfigurations. In Alamann and Bavarian law the word *maculare* was used in descriptions of this type of damage; meaning 'maculate', 'defile'. Formulations such as *aurem maculare* or *palpebris maculata* suggest that the crime described by their means was viewed as permanent damage to the body i.e., physical damage but equally as a violation of the victim's honour and dignity.

The aesthetic aspect also had an undoubted significance, something that was expressed literally in the quoted ascertainment as to the unsightly appearance of the mutilated ear. These regulations placed emphasis on the visual effects of the said deformation, which was to find full expression in one of the titles of Frisian law where there is talk of the deformation to the face being visible at a distance of twelve feet (approx. 3.5 metres): *Si ex percussione deformitas faciei illata fuerit, quae de XII pedum longitudine possit agnosci, quod wlitwam dicunt.*³¹⁸ We are also dealing here to a large degree with psycho-social effects arising

---

316 LRib, V, 2. *Si nasum excusserit, ut mucccare non possit* [...] *Si mucare praevalit* [...].
317 ERot, 50. *Si quis alii labrum absciderit* [...] *et si dentes apparuerint unus duo aut tres* [...]; 51. *Si quis alii dentem excusserit, qui in riso apparit* [...] *et si duo aut amplius fuerint in risu apparentis* [...].
318 Add. Sap., II, 26. Cf. A. Niederhellmann, *Arzt...*, pp. 288–289 as well as footnotes 8–16 (related literature): the Old Germanic word *wlitiwam*, composed of two words *wliti* and *wam*, meant the same as 'deformation of appearance'. The element *wliti* is connected with the Gothic *wlits* 'face', 'figure', the Old Frisian and Old English *wlite*, Old Swedish *wliti* 'face', 'appearance'. The second component of the term– *wam* – is linked to the Gothic *wammē* 'stain', 'flaw', 'damage', and also with the Old Icelandic *vamm* 'flaw', 'defect', 'disability', 'crippledom' and the Old English *wamm*, 'stain', 'imperfection', 'harm'; cf. Oliver, *Body...*, pp. 101-102: 'Stefan Jurasinski has recently pointed out that Old English *wamm* implies shame more than damage or disfigurement, citing the following parallel from *Maxims* 1 : (...) ("Word spreads, often they encompass one with *shameful [accusation]*, men speak of her insultingly, often her face becomes marred). It is perhaps relevant to the usage of wlitewamm in [Æthelberht] chapter 56 that public disgrace is linked in this passage, associatively, but perhaps causally, to embarrassing marks on one's face.") '; S. Jurasinski, *Germanism, Slapping nd the Cultural Context of Æthelberht's Code*, Philological Quarterly, vol. 80/1 (2002 for 2001), p. 61.

from the change in facial appearance, which stamped the victim of the said acts with the stigma of being different, ugly, comical or horrific in appearance.[319] The texts of the laws under analysis do not provide us with knowledge about the scale of the severity of the said facial disfigurations for an individual in social interactions. We may merely attempt to visualise it. However, it appears that the very existence of the cited regulations prove that early medieval Germanic societies recognised the problem of disfiguring damages to the body.

The next crime group represents a greater weighting from the point of view of their consequences. These are violations of the body or acts of violence which resulted in disfiguration and even damage to the organs of sensual perception and speech. The matter here concerns deafness and muteness, the loss of sight or head injuries resulting in oversensitivity to warmth and cold.[320] Important for us are obviously the descriptions of the individual cases of the results of bodily violation. Alamanns law (*Lex*) provides an interesting detail where talk refers to the case of the complete cutting out of the tongue, while in another to cutting out a half of it, so it was possible to understand what the victim said.[321] Worthy of emphasis is also the fact that in the law of the Thuringians, the Bavarians and the Frisians (*Add. Sap.*) there is talk of deafness being caused by a blow (wounding) to the head (here are used the words: *percutere, plagare*)[322], while in the laws of the Alamanns and Saxons the cutting off of the ear or ears is given as a cause of deafness.[323] Additionally, Frisian law took into consideration the case of a loss of speech (*ut mutus [...] efficiatur*) as a result of a blow to the head.[324]

---

319 Cf. J. Banaszkiewicz, *Złota ręka komesa Żelisława*, [in:] *Imagines potestatis. Rytuały, symbole i konteksty fabularne władzy zwierzchniej. Polska X–XV wieku*, ed. J. Banaszkiewicz, Warszawa 1994, pp. 232 ff.

320 *Add. Sap. Wul.*, II, 32. *Si homo ab alio ita in caput percussus fuerit, ut nec frigus nec calorem pro vulneris impatientia suffere possit* [...].

321 *LAl.*, LVII, 19. *Si autem lingua tota abscisa fuerit* [...]; 20. *Si autem media, ut aliquid intellegat quod loquatur* [...]; Cf. also: *PLSal*, XXIX, 15. *Si [quis] linguam alteri capulaverit, ut loqui non posit* [...]; *LSal*, (D, E) XLVIII, 12. *Si quis alterius linguam capulaverit, unde loquere non possit* [...].

322 *LThur*, 24. *Qui alium percusserit, ut surdus fiat* [...]; *LBaiw*, IV, 14. *Si eum sic plagaverit ut inde surdus fiat* [...]; *LFris*, XXII, 1. *Si quis alium per iram in capite percusserit, ut eum surdut efficiat* [...]; *Add. Sap.*, II, 18. *Si quis alium ita caput percusserit, ut surdus et mutus efficiatur* [...].

323 *PLAl*, VI, 2: *Si totum excusserit, aut placaverit, ut audire non possit* [...]; *LAl*, LVII, 8. *Si quis aliquis aurem alterius absciderit et non exsurdaverit* [...], 9. *Si autem sic absciderit profundo et eum exsurdaverit* [...]; *LSax*, 11, *Similiter de una aure vel ambabus, si surdus efficitur*.

324 *Add. Sap.*, II, 18: *Si quis alium ita caput percusserit, ut surdus et mutus efficiatur* [...].

In Alamann law (*Lex*) besides causing the eye ball to leave the socket there is also talk of a case of sight being affected, as well as the literal removal of sight from the eye (*visus tactus fuerit de oculo*) in such a way that it remains as if glass (*ita ut quasi vitro remaneat*).[325] In Frisian law the consequence of a blow to the eye is a loss of sight.[326] It is difficult to determine whether in these cases of more importance was a change in the appearance of the eye and the simultaneous disfiguration of the whole face or the disablement to sight.

Possibly, particularly in the case of the example of a provision from *Lex Alamannorum*, the matter concerned both outcomes of eye damage. It seems that this case illustrates a wider phenomenon involving the dual character of the damage to the face here described. A lot of the damage constituted its disfiguration while at the same time bringing about the deformation or loss of the given sense. Some of the quoted crimes could have had a humiliating character like the partial cutting off of the tongue, which distorted the victim's speech, creating a strange person differentiating himself in an unfavourable way. It was presumably similar to the knocking out of the front teeth, which resulted in presumably lisping or other speech defects. The disfiguring of the face and the impairment to the organs of perception are to a degree similar to each other although the weighting given to the damage that is the result of them sometimes differs noticeably.

Another type of consequence in relation to bodily damage that appears in the barbarian *leges* are the cases of loss or impairment to the limbs and motion ability. In Bavarian law there is talk of damage to the fingers resulting in an inability to bend them or to grip objects; damage of this nature made it impossible for the victim to bear arms.[327] In *Lex Alamannorum* we find a provision that talks of paralysis of the middle finger that resulted in the victim being unable to hold a shield and weapon.[328] Another regulation in Alamann law mentions a blow to the elbow, resulting in the victim being unable to carry objects or raise his hand to his mouth, and subsequently the paralysis of the entire arm, whereby the vic-

---

325 *LAl*, LVII, 13.
326 *LFris*, XXII, 45. *Si quis oculum alterius ita percusserit, ut eo ulterius videre non possit* [...].
327 This title refers to the cutting off of the fingers, after which there is a clause on the paralysis caused – *LBaiw*, IV, 11. *Et si non fuerint abscisi, et est mancus stat rectus, ut non possit plicare, hoc inpedimentum est ad arma baiolare, maior est compositio quam de absciso, tertiam partem supra addet.*
328 *LAl*, LVII, 53. *Si quis autem longissimum digitum sic placaverit, ut exinde mancus sit, ita ut conplicare non possit aut scutum prendere aut arma in terra propter illum recipere, 12 solidos componat.*

tim 'could do nothing with it'.³²⁹ *Pactus legis Alamannorum* in turn talks of damage to the leg, which resulted in the victim being only able to walk in his fields with the aid of a crutch.³³⁰ Therefore, we are here dealing with debilitating forms of damage or even those that make it impossible for actions needed to fight or work to be carried out as equally an inability to participate in daily activities. A specific type of crime effect is that connected with damage to the male genitals. For in *Pactus Lex Salica* as well as in Alamann law we have regulations that talk of damage to the penis resulting, one may presume, in an inability to have an erection (*unde mancus sit; ut virilia non tollat*).³³¹ It follows to add that a crime of this type represented a separate instance to total castration. Analysis of the contents of the provisions that talk of the consequences of damage to the body show that the individual barbarian codifiers understood the object of a crime to be not simply the body as such but also various functions, whose workings were possible thanks to the existence of the sensory organs and the limbs and their parts (usually the hand and its fingers). They included here, for example, sight and hearing, the ability to grip and move around, as well as the question of fertility. In addition the creators of these provisions had in mind the functioning of the smallest parts of the body, such as the eyelids, mouth and nose. Another subject of interest to the creators of the law was the appearance of a man's body, and first and foremost his face and its various components.

On the basis of the regulations on damage to the face one may ascertain that those who wrote down the law had a certain specified conception as to what its proper i.e., untouched state should be, and also about the values that the appearance of its constituent parts were to have. This can be seen especially well upon the example of the face as an object of crime, for the creators of the various regulations and provisions clearly saw a link between the physical state of this part of the body and a man's dignity and also honour. Lasting damage even to the smallest parts of the body such as the eyelids, eye brows, lips and all the more so the ear or nose constituted not simply an act of aggression but an attack on the body's social image. A man with a disfigured face (but also regarding other parts of the body) not only differed from the accepted norm as to appearance but the irreversible bodily damage inflicted could suggest that he had been subjected

---

329 *LAl*, LVII, 37. *Si enim in cubitu percussus fuerit, ita ut portare non possit nec ad os manu vertere* [...]; 38. *Si enim totus brachius mancus fuerit, ut nihil cum eo facere possit* [...].

330 *PLAl*, XI, 1. *Si quis alteri pedem truncaverit* [...]; 3. *Et si foris villam ambulare potuerit et in campo suo cum stelzia ambulare poterit* [...].

331 *PLSal*, XXIX, 17. *Si quis hominem ingenuum castrauerit aut uiriculam suam transcapulauerit, unde mancus sit, mallobergo gaferit, solidos C culpabilis iudicetur*; *LAl*, LVII, 59. *Si autem castraverit, ita ut virilia non tollat, 20 solidis componat.*

to maiming, which was meant to humiliate and lower him in society's eyes and was a sign of his disgrace, shame and dishonour.

The legal texts studied do not inform one of the circumstances under which the various forms of damage occurred but it seems that some of them were rather the result of unlawful mutilation than as a consequence of fights or brawls. The cutting off of both lips, the knocking out of both eyes, the cutting out of the tongue (in its entirety or partially) and also the amputation of particular sections of individual fingers as well as the removal of the testicles are examples of bodily violations which arose as a result of holding a man with the aim of his mutilation. In Salic law there is mention, for example, of the cutting off of three adjacent fingers (the middle, the ring finger and the little finger) with a single chop (*Si quis [alteri] sequentes uero digitos, is est tres si pariter in unum ictum excusserit [...]*).[332] While the cutting off of individual fingers was treated separately. The difference between the amputation of several fingers at the same time and individual severing is also expressed in the different values for damages.[333] They were, consequently, two separate categories of crime. Most likely it follows to suppose that the cutting off of three fingers in a single instance took place during battle, while the cutting off of individual fingers could have been the result of mutilation or torture.

The descriptions of bodily damage presented are extremely significant from the viewpoint of research into the perception of the body as an object of crime on the part of those writing down barbarian law. In provisions of this type the individual parts of the body were not perceived as an organic substance but first and foremost as 'objects' important from the viewpoint of the individual's participation in social life. For in them the matter concerns the functions connected with communication and perception (the ability to speak, listen, see), the carrying out of various activities connected with work, fighting, everyday life, as well

---

332 *PLSal*, XXIX, 7. The sentence may be translated as follows: 'While if someone cuts off the following fingers of another, that is three, to an equal degree in a single chop [...]'. This point occurs after the provision for the index finger (point 6), while about the cutting off individually of the middle, ring and little finger there is reference in point 9 of the quoted title. Point 7 may, therefore, be interpreted as concerning the middle, the ring and the little finger and not three fingers in general.

333 *PLSal*, XXIX, 4–9. The arrangements for cutting off individual fingers in *solidi*: the thumb – 50, the index finger – 35 *solidi*, the middle finger – 15, the ring finger – 9, the little finger – 15. Compenstion rates for cutting off: three adjacent fingers - 45; two adjacent fingers – 35 *solidi*. We can observe a similar depiction in *Lex Frisionum* (XXII, 28–33), where in 5 consecutive points there is described the amputation of each of the five fingers individually (from the thumb to the little finger), and then the case of severing them all. In point 34 there is talk of the cutting off of a fingerless hand.

as a man's appearance. The varied forms of damage disfiguring or completely eliminating the said functions and worsening the features of appearance indicate, to some degree through negation, the value bestowed on the elements of the body outlined.

The placing of descriptions of phenomena of this type within the texts of the *leges*, caused the parts of the body that appeared in them to be included within varied contexts, ones that exceeded the biological and organic features of man. Concrete parts of the body, and more strictly speaking their state (capability – incapability, integrality – mutilation) were consequently viewed from the point of view of the individual's everyday existence as well as the conditions of their functioning within society. In the provisions of the *leges* researched the eye, ear, tongue and hand were viewed in connection with their organic functions – sight, hearing, speech, ability to move and grip, which to a large degree determined a man's situation in life. Damage to or the complete depriving of a man of bodily functions of this type decided on a man's quality of life broadly understood, and in certain cases decided on the possibility for biological survival. Disfiguration of the face (scars, cutting off things, puncturing) or other forms of disfiguration (e.g., amputation of the fingers) surely lowered the position of the individual in the group, as a change in physical appearance could result in a worsening of the individual's social perception leading to marginalisation and alienation.

* * *

The questions presented – the cataloging of parts of the body and the crimes against their integrality appearing within the *leges* provisions – constitute a mutually related set of problems. For the provisions of the *leges* show each time the specific relationship of these elements (the bodily object – the crime). From the viewpoint of the subject of this study they are of a fundamental significance. However, it follows to equally note that the influence of a given act on a person's body was in the texts the subject of legal evaluation in relation to its damage to the victim. In the regulations researched we have therefore also a second relationship: between bodily harm and its monetary equivalent. In the case of each crime the law makers designated a clearly defined amount for compensation. The relation of crime – compensation constituted an important addition to the elementary relationship: body parts – the crime. As we shall see further on in the chapter the said equivalence of bodily harm allows one to view differently the cataloguing of body parts and crimes herein discussed.

Compensatory Systems and the Problem of Body Part Evaluation

Within the research conducted into those provisions covering crimes against bodily inviolability and integrality the problem of the relations between the character and range of damage experienced by the victim and the level of compensation that was to be paid by the perpetrator responsible for the said act has only very occasionally been touched upon. This question has at times interested historians of law as a part of investigations into certain types of bodily harm. These studies were of a fragmentary character and of a probing nature, usually covering selected crimes or individual parts of the body.[334] L. Oliver presents an interesting proposition for a holistic analysis of the system of compensation adopted in the *Laws of Æthelberht*.[335] This researcher has investigated in detail the assembly of payments designated in relation to damages inflicted on concrete parts of the body from the point of view of their objective severity, determined by the use of contemporary physiological and anatomical knowledge (e.g. the influence of the cutting off of the auricle (external ear) on the ability to hear). In the light of this data she underlines that the rate of payment resulted from an evaluation of the influence of the effects of crimes on mobility or the victim's perceptive abilities as well as also from the debasing nature of the damage.

The value of the compensation is treated by the author as an expression of the codifiers' decisions, ones dictated either by their knowledge and experience on the functioning of the body and the effects of various forms of damage, or by the criterion of visibility (or non-visibility) of the said violations. The author follows the numerical relations between the amounts paid for the particular crimes and also their relation according to the value of the *wergeld*. Oliver compares the rates of payment for various damages to the same parts of the body (e.g., the piercing and nicking of the nose – 9 and 6 shillings) or the functions (disfigured speech and its total loss – 12 and 25 shillings). She considers in connection with this that the damages resulting in a loss in perception functioning (e.g., hearing) were connected with higher rates of compensation than was the case for the loss

---

334 Cf. W. E. Wilda, *op. cit.*, pp. 768 ff., who compared and contrasted the cost of compensation for the cutting off of a hand and fingers in 20 legal codes (both Swedish, Norwegian and Danish from the 12th-13th century in early medieval *leges barbarorum*, and within Anglo-Saxon law); L. Günther, *op. cit.*, pp. 58, ff., has used the term 'compensatory system' and has researched the phenomenon in various Scandinavian codes and in *leges*; he has described them on the basis of the link between certain types of crime (bloody woundings, 'dry' violations, breakages) with the amounts of compensation; cf. also K. O. Scherner, *Kompositionensystem*, [in:] *HRG*, vol. II, column. 995.
335 L. Oliver, *Beginnings...*, pp. 99–105.

of an organ of perception (e.g., an ear). The difference in the amount of damages was to reflect the difference in the degree of severity for a given type of violation. She similarly interprets those cases in which the factor deciding on the rate and amount of compensation was whether the violation remained visible or invisible. The knocking out of 4 front teeth, which from a practical point of view is less severe than in the case of molars (enabling the grinding of food) was connected with a higher fine, for the former constitute a visible part of the body while the latter not (respectively 6 shillings and 1 shilling in compensation).

The referred to observations suggest that the amounts of fines entered in the *Laws of Æthelberht* constituted a certain system (although the author does not use such a concept) as well as being designated in accordance with certain principles (physiological effects, visual effects). One may gain the impression that the value of individual fines ascribed to subsequent cases of crime when compared to the nature of the damage to the body - that is, their effect - resulted from objective factors. The research also compares the list of forms of damage and bodily violation according to the rates of compensation, of which there were 22.[336] She does not, however, analyse this data in relation to the differences or similar character of these crime groups, even though some of these cover a dozen or so items. This has meant that certain features of the system of tariffs and fines has remained beyond the author's field of observation.

In her latest book (*The Body Legal in Barbarian Law*) L. Oliver encompasses the above interpretative method for the laws of Germanic peoples inhabiting other regions of western Europe. In this work she develops the questions she had earlier examined in relation to Anglo-Saxon law. Namely she advances the thesis that in barbarian *leges* the damages for forms of bodily harm that were more visible could have been higher than for those that did not immediately catch the eye.[337] Oliver cites regulations from various Germanic codes which could constitute examples of the two solutions advanced by her i.e., the open and hidden increase in the compensation rates (in relation to the 'ordinary' rate) for visible forms of damage. This thesis as to the existence of open additions (e.g., in the *Laws of Alfred*, King of Wessex, *Lex Alamannorum,* the Burgundian *Liber Constituionum, Edictum Rothari,* the *Laws of Æthelberht*)[338] is convinc-

---

336 *Ibidem*, pp. 195–198 (*Appendix II*).
337 Oliver, *Body*..., p. 165: 'In legal terms, these excessive fines represents punitive reparation – either overt and covert – added to the compensatory charges assigned for actual injury'.
338 Oliver, *Body*..., pp. 166-167, the codification of Alfred (*Ælfred*, 45 i 66, 1, [in:] *Die Gesetze der Anglesachsen*, vol. 1, Halle 1897) 'requires twofold compensation for a facial wound below the hairline'; Alamann Law (*LAl*, LXIV ,2) foresaw 'a twofold fine for causing a bruise to the face not covered by hair or bread'; the matter was similar in

ingly documented. Less obvious are the notes on regulations where the said increase was to have occurred in a hidden way.[339]

The existence in certain codifications of even only the first solution justifies the author's thesis '(...) no territory in barbarian Germania overlooks the importance of physical appearance'. Valid equally is her subsequent conclusion: 'Although the concept of honour-price is never directly addressed, the barbarian laws imply that shaming an opponent with deliberate disfigurement must be suitably recompensed.'[340] This view does not, however, substantiate Oliver's thesis as to the existence within the researched legal system of two separate components for the compensatory amounts: (1) 'punitive damages' for visible damage to the body, which was to be added to (2) 'purely compensatory damages' for anatomical losses.[341]

More accurate appears the thought that within Germanic law (as opposed to Irish) the compensation quota was the determinant of the social evaluation of the severity of a crime without differentiation into an honorary aspect and anatomical damage. This results from the fact that each form of damage 'purely physical' in nature (particularly that which was visible) was treated simultaneously as honorary loss. This last thesis is also contrary to the conclusion repeated several times in her analyses of particular Germanic codifications; that in certain regulations the authors of the said ignored the anatomical and functional meaning of a given part of the body in establishing the compensatory rates for their damage.[342] Such cases appear to prove that the anatomical and functional aspect of

---

Burgundian Law (*LBurg*, XI, 2); in the oldest Lombard law (*ERot*, 46) 'injuries above the hairline are fined less than those below'; Kentish law (*Ætel*, 60) 'assigns two fines for disfigurement of the countenance (*wlitiwamme*); that for a greater defacement six shillings, and for a lesser, three'. We might well consider all rulings involving shearing facial hair to be insult rather injury payment.

339   Oliver, *Body*..., p. 168: 'clear instance of covert punitive charge assessed on top of compensatory damage: greater visibility of front teeth makes their loss an insult', e.g., *ERot*, 50-52 where there is indicated a 'higher value to a tooth that appears in smile', in turn in Alamann Law (*LAl*, LVII, 18) awarded 'an increased fine if cut lip allows the teeth to show, thereby making wound more obvious'.
340   L. Oliver, *Body*..., p. 171.
341   L. Oliver, *Body*..., p. 170.
342   Oliver, *Body*..., p. 82: 'The only offence which demands an equal fine in Æthelberht is striking off the great toe; even today, injury to the outer tabulum of the skull and loss of great toe do not seem to be vastly inequivalent injuries'; *ibidem*, footnote 23: 'Alfred's compensations are left out of consideration here, as his laws offer a bewilderingly inconsistent list in terms of physiological value which is difficult to take too seriously since the same fine is assigned, for example, to striking of an arm below the elbow and striking off the 'shooting finger' or forefinger. Fifteen shilling fines are due for (in or-

bodily damage was not treated as a single independent criterion in establishing the compensatory value allocated in a given case.

As we claimed above there are three basic elements in the regulations of interest to us: the description of the type of damage or violation, its bodily subject (the most often the designation of a given body part) as well as the value of the compensation i.e., a definite sum of money. Therefore the sets of rules on crimes against the body constituted specific fine tariffs imposed by the law for the committing of designated forms of damage and bodily inviolability. This form or arrangement of the said tariffs raises the question as to the relationship between the type of crime and the given part of the body on the one hand, and the amount of compensation on the other. This is in essence a question about what the sums for the fines were and in what way a contemporary researcher may interpret them. What was the main factor shaping their value – the type of violation or damage, the nature of the bodily object subjected to them, the circumstances in which the crimes were committed or also the product of these elements? With this is connected the question of the possible directions of interpretation for the said tariffs: do they reflect the valuation of the weighting of the crimes themselves, the subjective evaluation of the various body parts or both?

The Origin of the Compensatory System for Violation of the Human Body

In analysing the provisions of Germanic *leges* in their treatment of various forms of violation and damage to the body we face the question as to what the reason for setting specific amounts of compensation for individual crimes was. This matter only seemingly appears obvious in nature. For the principle itself that violation of a given body part brings with it the necessity for material compensation on the part of the crime's perpetrator is not an obvious concept. This is

---

der) an outer head wound; knocked-out canine tooth; slit cheek; arm cut off below elbow; forefinger off; 2nd (not great) toe off; piercing into scrotum; cut from shoulderbone; cut from rib.' p. 106: 'This analysis presents a clear distinction of purpose in these clauses. Whereas legislation in Bavaria, Frisia and Wessex approximates physiological worth, that in Salic Francia, Burgundy, and Wessex [probably: Visigoths – my comment] ignores any differentiation. All other territories, however, place greater importance on aesthetic than functional worth.'; pp. 153, 155: 'There is neither a physiological nor a social explanation for the Thuringian equation of all fingers, including the thumb. Striking off any digit requires payment of 1/6 *wergeld*, which is ridiculously high for the ring and middle finger. The Visigoths regulate according to sequence, with the forefinger most highly valued and the little finger least. These last two regions employ bad anatomy and bad social analysis, and their ruling thus represent bad law.'.

borne out by the fact that modern law does not apply such a principle.[343] What is more already in codifications from the late Middle Ages there do not appear regulations on damage for violation of particular parts of the body. In the early medieval laws of the Germanic peoples they functioned, however, as dogma and a cardinal principle. This makes the analysis of the problem of the mechanism for the creation of the said system of compensation all the more important.

In attempting to explain this basic problem we may adopt the following hypotheses. Firstly: we may start from the premise that damage to a person's body as a result of a crime required material compensation for the victim's loss. In detail the members of the body mentioned in the Germanic codes had an objective value resulting from their position in the human organism. Certain parts had consequently a special significance resulting from the role they performed in a human being's life. These certainly included: the eye, the hand and arm, the feet and legs, the genitals etc. A person deprived of any of these, or having experienced serious damage inflicted to the said was to encouter real, concrete and negative effects. For certain, the cutting off or damage to a single finger or toe or the knocking out of a tooth did not result in such severe effects. Hence the varied ranking given to these parts of the body.

The second explanation of the origin of the compensation system is as follows: violation of the body as a result of crime caused a feud i.e., a state of hostility between the victim and the perpetrator and their families. For every form of bodily damage constituted an insult to the victim; to preserve honour the damage should be avenged or equalled in another way.[344] To stop the animosity i.e., in order that the victim did not take revenge on the perpetrator, compensation was proposed relative to the type of harm inflicted. Compensation was therefore the price for not taking revenge.[345] However, this explanation, alt-

---

343 Cf. for example the provisions of modern Polish law: *Kodeks karny*, Warszawa–Bielsko-Biała 2008, (p. 72) art. 159 (concerning serious damage to the body such as: 'depriving a person of sight, hearing, speech, the ability to procreate') and art. 157 (§ 1 on the causing of violation of a body organ or a health disorder not outlined in art. 156 § 1).

344 Cf. J. Smith, *op. cit.*, p. 101; the author claims that the compensation amounts were '[...] far more than a payment for physical damage. It expessed in monetary terms the slight done to familial pride, reputation and integrity [...]'; these payments were compensation for loss and a form of reinstating social respect. The author bases her argumentation on early medieval Irish laws, in which there existed two separate payments for wounding: one constituted a compensation for the injury, while the other was a payment for 'a clearing of one's face', which was to eliminate the stain to one's honour (Cf. N. Mc Leod, *Parallel and paradox...*, pp. 52–53). This view is justified. However, we consider that in Germanic laws there did not appear the said division into two forms of payment, additionally this does not explain the differentiation in the values of the damages.

345 Cf. *ERot*, 73; cf. K. Modzelewski, *Barbarzyńska...*, pp. 119 ff.

hough having its basis in the form of the clauses in *Edictum Rothari* talking about the increase in the level of compensation for bodily violation to avoid feuds, raises doubts. For it seems that the principle of a 'ransom' for a refraining from revenge e.g., for a bloodless blow with a stick does not seem clear. The source of the 'price of buying out' a feud was in essence the 'evaluation' of the bodily damage.

The thesis that compensation represented ransom for the abandonment of feuding has also one more weak side. For how was it possible to differentiate the level of revenge in the case of two extremely similar crimes, e.g., between the cutting off of a third or two thirds of a middle finger? One may, obviously, contemplate the hypothesis that in a similar way to murder where revenge involved the killing of the perpetrator, then every form of bodily damage was 'settled' by the inflicting of the self same damage to the perpetrator's body. In such a case the fines would in essence be the price for buying the relevant body part from the criminal. Such an understanding has at least one weak side. In a situation where, for example, a freedman cut off the hand of a freeman, he would – in accordance with this principle – pay for 'his' hand, which would be 'worth' less than the one belonging to the freeman. This example shows that compensation could not have been the price of buying the body part of the crime's perpetrator, one analogical to that which was the subject of damage for the victim, but constituted compensation for the victim's loss.

Lisi Oliver has proposed a hypothesis on the way in which in court practice the tariffs for body damages were applied. These payment amounts were in her opinion to have constituted the highest compensatory amounts that could be applied. Germanic judges were to have taken into consideration the circumstances in which a crime was committed and when the situation justified it they could reduce the amount to be paid the victim. This conjecture the author bases on the regulations of Visigoth and Frisian law – the former states that in the case of a wounding of the nostril the judge may alone stipulate the amount, while the latter states that for the wounding of the hand of a harpist or goldsmith the compensation was to be a quarter higher than in the case of other people.[346] These regulations point, in her opinion, to a situation wherein the amounts of compensation could be determined by judges. The possibility of being able to lower compensatory rates was to have had a general application within the Germanic legal system. Oliver notes that the latter statement is merely speculation. This hypothesis, although certainly not impossible, is, however, difficult to verify upon the basis of the texts of the Germanic *leges* alone. We do not have sources

---

346 Oliver, *Body..*, p. 51, *LVis*, VI, 4, 3; *LFris*, (*Haec iudicia Wlemarus dictavit*, 10), pp. 90-91.

talking of the relationship of these regulations to court practice, something the author clearly underlines. None of the known Germanic codes possess regulations that speak of the possibility or means by which rates established in the tariffs for damages could be reduced or more broadly how their values could have been altered.

The system of fines clearly went beyond the logic of ransom for giving up on revenge. It seems that it established in fact the degree of responsibility for the individual violations to the body. A blood feud would not have given such a possibility. Therefore, the establishment of a system of damages constituted, as may be ascertained, a considerable breakthrough in the criminal history of the body. The ascribing of a specific value for compensation to individual parts of the body resulted in their objectivization, for each body fragment that had been incorporated into the system became an object defined by law. The system of damages became a network of values put, as if, on a man's body. Man's body was comprehended within a specific scheme constituting a form of defining its valorisation. Each part of the body listed in the rates for compensation was a part of the system and gained a specific place within it.

Tariff Payments for Bodily Injuries

In the provisions relating to crimes against the body, the prevailing principle is that for every act of this type, the perpetrator is supposed to pay out a fixed sum of compensation to the injured party. A certain sum of compensation was therefore prescribed for every injury to a particular body part. This regularity may suggest that the compensation tariffs reflected the 'value' of various human body members. But closer investigation of these tariffs may lead us to question this viewpoint. For if we rank the crimes they list in order of level of compensation, it turns out that in many cases many different injuries to various body parts were assigned the same amount of compensation. Analysis of the compensation tariffs for injuries to individual body parts shows that their level was not set separately for each crime. In many Germanic *leges*, we are dealing with a 'gradual' system of compensation rates of a threshold payment character. Anything from 6 to 19 'thresholds' of this kind appear in the investigated texts, regardless of how many crimes against the body are being taken into account. The lowest rate (6) was established in the *Lex Thuringorum* to cover 24 crimes. The highest number of 'thresholds' (19) appears in the *Lex Frisionum*, yet 80 crimes are defined here. In the other codes, the number of rates oscillates between these two extreme values. In the *Pactus legis Salicae*, there were 10 rates (for 36 crimes), while in the *Lex Salica*, there were 9 thresholds (for 20 crimes), in the *Lex Ribuaria*, there were 10 rates (for 22 crimes), in the *Lex Saxonum*, there were 13 (for

36 crimes), in the Visigothic laws, there were 10 (for 25 crimes), in the law of the Burgundians, there were 8 (for 11 crimes), in the *Pactus legis Alamannorum*, there were 12 (for 44 crimes), in the *Lex Alamannorum*, there were 12 thresholds (for 62 crimes), in the *Edictum Rothari*, there were 17 rates (for 48 crimes) and in the *Laws of Æthelberht*, there were 22 (for 62 crimes). Clearly, in all these cases the number of crimes was greater than the number of 'thresholds'.

This prompts the question as to what mechanism lay behind the creation of these threshold-based tariffs. Were they created in an evolutionary way through the addition of successive elements or by means of a single overall ruling. The second issue is whether these systems were created from the largest amounts with their gradually lowering or whether the opposite was the case and they were created from the crimes to which the lowest levels of compensations were applied and the scale of rates was extended 'upwards'.

The compensation tariffs arose, one might suppose, from a single overall ruling based on a carefully conceived plan that was approved in advance. The main reason for this is the fact that the Germanic codifiers usually treated crimes against the body (at least in relation to freemen) as an overall set. Often provisions relating to bodily injuries were set out in the order of appearance of the body part, from the head to the feet. This triggered a need to create overall compensation tariffs.

It seems that when the codifiers assigned compensation amounts, they began with those which were the highest. Let us investigate how this mechanism works in relation to selective instances. In the Lombard *Edictum Rothari*, the three highest compensation rates were expressed in the form of fractions of the *wergeld* for a free man: 2/3 for knocking out a second eye, 1/2 for knocking out one eye or amputating a nose, foot or hand and 1/4 of the *wergeld* for causing hand or foot paralysis and for bone fractures of the arm, which had yet to heal after a year. Such a method of assigning compensation payments clearly attests to the fact that *wergeld* formed the basis and starting point for determining their value. It was divided into smaller portions. The codifiers ascribed the sums which were the largest proportions of the *wergeld* (which itself fulfilled the role of a kind of scale) to those bodily injuries they regarded as most serious. The percentages of *wergeld* payable as compensation rates for less severe injuries were gradually decreased. In the *Pactus legis Salicae* the highest amount of compensation for complete castration (200 *solidi*) was equal to the level of *wergeld* for a freeman. The next compensation rate amounted to 100 *solidi*, i.e. half of this sum. This was prescribed in the case of foot, hand, ear or nose amputations or in cases where an eye had been knocked out or impotence had been caused. It therefore seems that in this case as well the compensation rate was prescribed by commencing with the highest rate and then dividing the *wergeld* into smaller sums.

Another example of this kind of mechanism in action is provided by the provisions contained in the *Lex Saxonum*. In these, a distinctive sequence of 'thresholds' appears for compensation payments prescribed for powerful and wealthy Saxons (noblemen) – 1440, 720, 620, 360, 310, 160, 155, 120, 80, 60 and 40 *solidi*. The second rate (720) amounts to half of the first (1440), while the fourth (360) is half of the second. A similar model recurs several times, e.g. 720, 360 and 180 or 620, 310, 155. It should be added that the highest compensation – 1440 *solidi* (prescribed in cases when both eyes were knocked out, or both ears or testicles cut off) was equal to the *wergeld* for an nobleman (adaling). The threshold system in the *Lex Saxonum* was based on four sequences of interlinked numbers: a) 1440, 720, 360, 180, b) 620, 310, 155, c) 240, 120, 60, d) 160, 80, 40. It is notable that the lowest rate – 40 *solidi* (for the amputation of a middle or ring finger joint) – amounts to 1/36 of the highest compensation amount (i.e., 1440 *solidi*). This prompts the assumption that in the law of the Saxons the 'threshold' system could have been constructed from 'the top down' rather than from 'the bottom up'. On the other hand, however, not all the cited rates are multiples of the number 40 (e.g., 60, 180, 310, 620). So it did not fulfill the role of an initial value which was then scaled up. Furthermore, as we already know, the highest compensation sum was equal to the *wergeld* for an adaling, which indicates that consecutive 'thresholds' were stipulated from 'the top down' by dividing them into fractions.

In the *Lex Saxonum* we are undoubtedly dealing with an overall system of compensation amounts rather than a mere random set. The levels of individual 'thresholds' were interlinked through the relationships indicated above. This was clearly a complex and carefully conceived structure. A compensation system of this kind certainly could not have come about by chance. It also seems to be extremely unlikely that this system evolved. It can be assumed that it was immediately constructed as a complete entity. It was a system modelled on numerical fractions. In Saxon law, the systemic nature of compensation tariffs is easily perceptible. Its numerical structure is evident and also closed: it is opened by the numerals 1440 and 720 and closes with 80 and 40. In both of these pairs, the sums remain in the same proportion to each other, while all of them are divisible by 40. A somewhat similar yet simpler 'threshold' system was adopted in the *Lex Thuringorum*. These contained six rates – 300, 150, 90, 33 1/2 and 30 *solidi*. The first and last sum are in a ratio of 10:1 with each other. However, other numerical relations appear here – between the rates of 300, 150 and 50 or 90 and 30 *solidi*. There is also an apparent tendency to combine them into groups of bodily injuries which were prescribed the level of composition – titles 4, 6. – 30 *solidi*; titles 8, 10.-11. – 90 *solidi*; titles 12, 14, 15-16, 18 – 300 *solidi*, titles 19-21 – 33 1/3 *solidi*, but titles 22 – 30, titles 23 - 50. and title 23 – 300 *solidi*.[347]

The presented compensation rate systems were clearly complex systems that were often based on recurring numerical ratios. It would appear that their creation involved assigning successively lower payment 'thresholds'. Admittedly, in the case of a large number of rates, the system of interdependencies between them became complicated. It also appears that this phenomenon forced codifiers to adjust compensation amounts more directly to the nature of individual crimes. One example of a mechanism of this kind for creating a compensation system is the *Lex Frisionum*.[348] As we mentioned, this laid out 19 composition rates. The way these were calculated merely serves to highlight the difference between the Frisian and, for example, the Saxon compensation tariff system. The rates are as follows: equivalent value to the *wergeld* (53 *solidi* and 1 *denarius*), 45, 41, half of the *wergeld* (26 2/3 sol.), 24, 20 2/3, 18, 15, 13 1/3, 12, 8, 7, 6, 5, 4, 3, 2, 1 and 1/2 a *solidus*. This is not simply a matter of there being a difference in the number of 'thresholds', it also concerns the relationships between amounts. The Frisian system of rates is less gradual and more 'fluid'. This is particularly evident in the divisions occurring between 8 and 1/2 a *solidus*. The differences between amounts are generally less here and, above all, in many cases they are not in a relationship of multiples with each other. Yet, even in this system, that kind of relationship is evident, e.g., 53 *solidi* and 1 *denarius*, and 26 2/3 and 13 1/3 *solidi*, or 24, 18, 12 and 6, or 45, 15 and 5 *solidi*.

Regardless of whether the numerical relationships are more or less obvious, especially when they take the form of sequences composed of multiples of a

---

347 Crimes according to the order adopted in the *Lex Thuringorum*: 4. a blow, 6. bloody wound, 8. any bone fracture, 10. torso puncture, 11. leg or arm puncture, 12. knocking out of one or both eyes, 14. nose, ear or tongue amputations, 15. hand or foot amputation, paralysis of hand or foot, 16. amputation of one or both testicles, 18. penis amputation, 19. thumb amputation, 20. index or middle finger amputation, 21. ring or small finger amputation, 22. toe amputation, 23. facial disfigurement, 24. blow causing deafness. Compensation values for noblemen.

348 H. Siems, *Studien...*, p. 222 ff., claims that the compensation system in the *Lex Frisionum* was largely constructed on the basis of the level of *wergelds*, i.e., of compensation (*Buße*) payable on the murder of a nobleman, freeman and half-free man (*litus*). Composition levels assigned in the case of the victim losing an eye, hand, foot and tongue were clearly linked to the level of *wergelds*. Besides this, in the author's opinion, a duodecimal system appeared in the Frisian law's compensatory system (this is exemplified by the following frequently recurring numerals: 4, 12, 2, 6, 24, 3, 8, 1 and 18 *solidi*); this system played a crucial role, together with the value of the *wergeld*, in the shaping of the compensatory value system. Siems also asserts that the decimal system was incomparably less important in the *Lex Frisionum* than the duodecimal system.

given number or divisible individual rates, the same model for constructing compensation tariffs applied in all the investigated codes. This was a sequence of amounts ordered from the highest to the lowest. This was a closed entirety which embraced every case of bodily injury listed in any given law code. The highest compensation rate in all the codifications was always in a clear numerical relationship with the amount of *wergeld* for a free man, being equivalent to or half of that sum, or in exceptional cases another proportion of that sum.

The Arbitrary and Comprehensive Nature of the Compensation Systems

Although the 'threshold' compensation rate systems identifiable in the investigated law codes manifest themselves as complex structures governed by their own rule systems, in fact they merely act as a complement to the list of crimes against the body contained in those codifications. Creating compensation tariffs, as we have claimed above, did not involve the definition of the level of compensation for every single crime against the body. This prompts the following question: How were individual compensation rates combined with particular cases of bodily violation? Undoubtedly, the mechanism for creating compensation tariffs in every law index required the creation of:

1) lists of separate body parts
2) a catalogue of different ways in which they could be violated
3) a system of compensation rates

The next issue emerges at this juncture: In what order did these three elements occur? The logic dictating the creation of legal regulations would indicate that the codifiers must have first compiled a list of body parts that were important to them and types of injury to these body parts. Next, they would have assigned compensation rates. In order to avoid doing this separately for each type of crime, they must have first divided violations of the particular body parts appearing on the list into groups. The next step was to establish their order according to the criterion of the relative gravity of each crime – from the most serious to the mildest. The codifiers assigned a specific compensation rate to each of these groups or categories of bodily violations. However, rather than assessing every single group of crimes, they were guided by the relationships between them, making sure that the arrangement of rates created a coherent system.

The crucial issues in this procedure were the criteria according to which given cases of bodily violation were combined into groups and also the principles employed for the creation of a hierarchy of importance for crime categories created in this manner. The basis for the creation of a crime group was, on the one hand, the location of a given body part in the value hierarchy adopted by the codifiers, and on the other, the severity of the injury to this part. According to the hypothesis posited at the beginning of our reflections, this was a 'natural' hierarchy of importance for body members, which stemmed from their role in the organism and the function they fulfilled in human life. An interpretative difficulty connected to this hypothesis resides in the fact that in legislative practice, this 'natural' hierarchy is not of a universal nature. The issue of the principles employed for establishing hierarchies of crime groups is also complex. The types of body part violation that were regarded as being similar in these respects were placed in one category. However, as we will make clear, the crimes placed in this same group do not always appear to be comparable to the modern reader. The third issue that appears in our reflections is the manner in which compensation rate levels assigned to particular crime groups were prescribed.

In order to better understand the issues raised here, let us examine some compensation tariffs payable for bodily injuries contained in the investigated law codes of the Germanic peoples. When analysing these examples, we will attempt to explain the following: the mechanisms behind the combination of the crimes being discussed here into groups, as well as that which established a hierarchy of importance for them within the framework of a given compensation system and that which stipulated the levels of given compensation 'thresholds'.

Cases in which a given compensation rate was prescribed for only one crime are rare. Yet even in these cases, it was not usually an amount resulting from an individual estimation of the 'value' of the body part in question. In the *Lex Baiwariorum*, in a set of 22 crimes against the body there is only a single case of a particular rate being ascribed to an injury to one body part (thumb amputation – 16 *solidi*). In the *Lex Frisionum*, 4 such cases appear for 80 crimes. In the *Edictum Rothari*, there were in fact 6 such situations for 44 crimes, and in two of these the compensation rate was assessed as a fraction of the *wergeld*, which means that it was matched to the 'threshold' model. We find a relatively large amount of crimes with an individual compensation rate in the *Lex Saxonum* – 4 out of 36. However, three of these were sums which were entered into sequences of numbers which were interlinked in a relationship based on multiples. A similar tendency appeared in other law indexes.

In the vast majority of cases fines of this self same amount were ascribed to many different crimes. There are so many examples of such a situation in Germanic *leges* that we shall mention only a few of them. The regulations of *Lex Alamannorum* are good material to illustrate this phenomenon. This is visible both in the case of the high as well as the low amounts of compensation. Nine different crimes are included in the quota of 40 *solidi*: cleaving open the skull, cutting off the ear (with resulting deafness), knocking out an eye, cutting off the nose, tongue and arm from the shoulder, testicles and feet as well as puncturing the innards. We have here nine parts of the body and four different types of injury. The rate of compensation, 12 *solidi*, was adopted in reference to 13 crimes: cutting off an ear (without resulting deafness), the tip of the nose, the thumb, the little finger, deforming the lower eyelid, puncturing the lower lip, knocking out two lower front teeth, shaving off hair, causing paralysis of the middle finger, damaging internal organs, puncturing both thighs and damaging the knee (causing lameness). Here are twelve different body parts and nine forms of injury. Finally the 6 *solidi* rate was applied to 12 crimes: breaking a bone (unspecified), deforming the upper eyelid and upper lip, cutting the nose, neck, arm above the elbow and side, knocking out two back teeth, cutting off the tip of the thumb, the middle finger and the little finger. Here there appear eleven body parts and 5 types of violation.

In total 35 out of the 62 crimes against the body of a freeman that are listed in *Lex Alamannorum* are ascribed to these three rates. In the case of each of the said 'thresholds' we can observe the same situation – the subordination to them of many different, often extremely varied in relation to each other, forms of damage to various body parts. The principles upon which the level of damages was determined for the various violations are not clear. Why, for example, was there determined the same rate of compensation for the cutting off of the nose or the knocking out of an eye and for the causing of paralysis in the middle finger or wounding internal organs? In as far as it is understandable that, for example, the loss of an eye was far more severe than the deforming of the upper lip, it is not clear why the compensation for the former was 40 and for the latter 6 *solidi*. On what basis were amounts of compensation estimated and in what way was there determined between them a numerical proportion (40 to 6). Alamann Law, in a similar way to other codes, provides no answer to this question. The codifiers did not give reasons for their decisions with regard to the amounts of compensation and the proportion between them.

Individual codes differ amongst themselves in their way of fixing the said rates. The special regulating of damage rates did not mean, however, arbitrariness and the lack of a link with corporeal reality. For example, in *Lex Saxonum* the damages for cutting off a hand were 720, a thumb – 360, while a big toe – 180 *solidi*, i.e. 1/2, 1/4 and 1/8 of a nobleman's *wergeld*. The gradation of amounts is, so it seems, in accordance with the role played by the cited members within the structure of the body. The hand is more important that the thumb, while the thumb more important than the big toe. Yet where did they get the amounts of 720, 360 and 180 *solidi* from? The fact that in the case of the cutting off of a hand constituted a half, the thumb a quarter while the big toe an eighth of a nobleman's *wergeld* does little to explain matters. For what was the justification for such proportions? In *Edictum Rothari* these relations are somewhat differently shaped. Admittedly the cutting off of somebody's hand was connected with the necessity to pay damages constituting a half of the *wergeld* of a free man, the damages paid for cutting off a thumb were a sixth of a *wergeld* (and not like for the Saxons a quarter); while for the analogical damage to the big toe – 16 *solidi*, which is a half of the amount for a thumb (i.e., $1/12^{th}$ and not $1/8^{th}$ of a *wergeld*). In turn, in *Pactus legis Salicae* these damages were 100, 50 and 50 *solidi* (i.e. a half, 1/4 and 1/4 of the *wergeld* of a free man). The proportion between the compensation for the cutting off of a hand and a thumb is therefore the same as it is for the Saxons, yet the big toe was 'valued' by the Salic Franks as equal to the thumb (differently than for the Lombards and the Saxons).

In *Lex Frisionum* compensation for the cutting off of these three body parts was 45, 13 1/3 snd 8 *solidi*. The *wergeld* of a free Frisian was 53 1/3 *solidi*. Damages constituted therefore not a half (as was the case in *Lex Saxonum*, *Edictum Rothari* and *Pactus legis Salicae*), but around $4/5^{th}$ of a *wergeld*, while compensation for analogical damage to the thumb was exactly a quarter of the *wergeld* (like for the Saxons). The proportion between the amounts of compensation for damage to the hand and thumb (more or less 3 to 1) resembles only those that appear in *Edictum Rothari* (a half and a sixth of the *wergeld*) and were different than for the Saxons and the Franks. The 'value' of the big toe of a free Frisian was about two thirds of that of a thumb. It was, therefore, different than in the other three codes.

We have here touched on only a few of the numerous examples of differences in the proportions between compensation for the same damage inflicted on the same body parts as they may be perceived when reviewing the various legal codes. This phenomenon shows that in each codification the value of the compensation and the proportions between each type was determined in a different way. There existed, obviously, also similarities, for example, in *Edictum Rothari*, in *Pactus legis Salicae*, in *Lex Salica*, in the laws of the Bavarians, Thuringians and Saxons the relations between the values of compensation for knocking out an eye, cutting off a hand and foot were the same, constituting 1 to 1. In the majority of these they constituted a half of the *wergeld* of a freeman. This fact did not mean, however, that the establishment of these amounts was conducted on the basis of some objective, universal principle.[349]

---

349 Cf. L. Oliver, *Body*..., p. 137-140 proposes a thesis that in the Germanic laws functioned pattern of 'Matthew limbs' (reference to the *Gospel of Matthew*, 18, 8-9), that is that value of compensations for damages of a hand, a foot and an eye were in most cases equal (see Table 5.1 Assessment of 'Matthew limbs'). However, she remarks that these

The gradation of the amounts of fines was shaped by two factors. On the one hand, the level of the individual damages was in accordance with 'the natural hierarchy' of the individual parts of the body, resulting from their function in the organism and the role fulfilled by them in the existence of man. For damage to body parts considered more important in this regard (e.g., the eye, hand, foot) a higher level of compensation was set than was for those of a lower ranking (e.g., toes, teeth etc.). However, one should emphasize that in some codes the order of the various members in relation to their importance was different. The said 'natural hierarchy' did not, therefore, have a universal character. For example: in some codes the male genitals were given a higher position (e.g., the codification of the Salic Franks) while in others the eye (e.g., *Edictum Rothari*). Both these body parts belonged to those members which are generally ascribed the highest levels of compensation in the various codes researched, but in particular codes they occupied different places within the hierarchy. The case was similar with other members.

The establishment of a hierarchy of body parts did not determine the value of concrete forms of compensation for given damage and violation. As we have already ascertained in many legal codes we are dealing with a clear relationship, namely of the highest values of compensation and the value of the *wergeld*. This observation does not result in a better understanding of the mechanism of 'estimating' the value of compensation. For it is unknown either how the amount of blood money was established, or what decided as to the numerical proportion between this amount and the value of a given set of discretionary damages.

One can obviously not exclude the existence of some objective practical factors (economic, social or other) which conditioned the level of a freeman's *wergeld*.[350] We do not know, however, whether blood money was to cover material loss and damage, such as would be experienced by the family of the deceased (and if so what and within what scope), whether it was to constitute 'social' security for the widow and orphaned children, or whether it had other justifications – e.g., to counteract bloody revenge and increasing the spiral of violence. For it could equally have constituted an indicator of the social position of a given individual. The second unknown involves the proportions of the *wergeld* and

---

laws 'equate the value of nose and ear with that of an eye'. She points out as well that 'it would be a mistake to connect the biblical association with the laws'. Although an observation concerning the equation of values of 'Matthew limbs' is right it does not explain the reason of using it.

350 Here the fact stands out that the value of the *wergeld* and damages was expressed in money and not by means of other units of value, e.g., heads of cattle, horses, land, slaves, valuables. This suggests that the sums were not directly associated with usable or production values.

the discretionary damages. However, even if it were possible to indicate some kind of criterion for determining these relations, it would not be able to explain fully the question of the origin of the value of the damages, as not all of them were a derivative of blood money.

An important argument for the sums of fines not being a reflection of some 'objective valuation' of body parts determined as a result of their degree of violation, is seen in the tariffs researched by the practice involving accreditation of various crimes against various parts of the body to the self same rate of compensation. This fact indicates that the creation of the system of damages was arbitrary in nature. It involved the grouping together of regulations on specific parts of the body and the ascribing to them of amounts of compensation. We have a lot of examples of such practices in Germanic legal codes. Individual crimes were classified to strictly determined amounts of compensation. The payment tariffs and rates contained within the codes under investigation did not foresee any possibility for negotiation as to the rate of compensation. These amounts were not connected, therefore, with an individual situation of a concrete crime victim or with individual cases.

Individual body parts or fragments were viewed by those writting down the law as objects of a standard or rather intentionally standardised quality. In the *leges* we are dealing therefore with a generalised, impersonal hand, eye or leg. Secondly, there is also here mention of crimes that have been taken as set model acts – cutting off (severing), breakages or wounding as a generalised concept and not as a reflection of an actual individual act of this type. In other words, in an individual regulation there is presented a model crime involving the damage inflicted on a standard 'averaged' body fragment. If we are to look at the research material in this way we can understand better the nature of the compensatory system. The means by which their value was established reflected that very understanding of the crimes outlined and the perception of the body fragments which they related to.

The interpretation of payment rates as presented in the case of various types of body damage and violation should incorporate one more important factor. The compositional amounts which appear in the codes under consideration resulted from an estimation of the 'value' or extent of the damage that resulted from the injury or loss of a given body part or organ. They did not, however, reflect the real value or potential advantage that could have resulted from being in possession of an undamaged, fully functioning hand, eye or other body element. They cannot, therefore, be interpreted as an expression of their 'positive' value, but merely as an estimation of the effects resulting from their loss or disablement. The said list of crimes and compensation amounts may, therefore, be interpreted as more or less a complete register of deviations from the condition of the individual body fragments as considered to be the normal state of a man's bodily shell as viewed by the compilers of the *leges*. This was, consequently, a 'negative' valuation of the human body.

The above comments are of considerable significance in determining the question as to whether on their basis one may determine the means of valuating body parts or fragments on the part of those codifying the various legal codes. For certain the amounts which are mentioned in these regulations cannot be directly treated as an expression of body valorization. The ruling which proclaims that compensation for the cutting off of a hand should be x *solidi* cannot be interpreted to mean that the creators of this norm evaluated the objective value of the said body part in accordance with the value of the actual amount cited. The values of damages attributed to a given body part was different depending on the type and degree of its violation; for example, for the cutting off of a nose and its perforation two different rates of compensation were determined. This is all the more visible when we take into consideration such cases as the wounding of the face and the slapping of the face, or the cutting off of hair and the pulling of hair. The compensation rates allocated for this type of violation do not constitute an estimation as to the value of the face or hair. They determine the value of the degree of violation to the bodily integrality or inviolability of an individual.

Upon the basis of the amount systems for compensation outlined in the *leges* studied we may state only that their creators considered certain body fragments (members or organs) to be more important, while others less so. The most valued included: the eye, male genitals, hand, foot, nose, ear, tongue, leg; another group with regard to value covered: fingers, the arm, the torso, the skull and the scalp; of noticeably low valuation were teeth, hair, lips, eyelids, limb bones, toes, skin anywhere bar the head, the cheeks. The position of a given body part, which resulted from the amount of compensation allocated in the case of its loss or damage, was determined in its relationship to other bodily elements as well as it relation to the *wergeld*, i.e., money paid for murder. The amounts of fines did not have, therefore, an independent absolute meaning. The valorization of a body part was relative in nature. The amount of compensation was, however, shaped by the nature of the damages and violation inflicted.

Chapter Four

The Perception of the Human Body and Social Differentiation amongst Germanic Peoples

In the previous chapter we dealt with the differences in the rate of fines indicated for damages to particular body parts or bodily objects as well as the differences that existed in this case amongst the various legal codes. The regulations for a particular crime victim group that of freemen has served us as material in these analyses. The problem of crimes against the body and the damages ascribed to them needs, however, to be examined within the context of the differences in the socio-legal status of various groups or categories of individuals appearing within the particular barbarian *leges*. Hence at present the subject of our investigations will be the problem of differences within the rates of payment (compensation and *wergeld*) established in the case of these same types of crime for particular social groups. This is connected with a series of issues which demand detailed analysis. The first question being the reasons for the said difference. The second problem is the way of presenting the social structure of the given people and the collection of body parts subject to crime as well as the relation of these two elements.[351] Another matter is the connection between a man's body as an object of crime and the concept of social order contained within the codes under investigation.

Before we move onto these matters we shall clarify in what way the said social differentiations in compenstion manifested themselves. Expressing matters in the most general fashion – in the case of committing crimes against the body i.e., violation of inviolability of bodily integrity, the perpetrator was to pay the victim (or his legal guardian) a money payment (or its equivalent) determined by the law, the value of which depended on the social position of the victim and his affiliation to one of the social groups with a different legal status. In order to

---

351 The problem of the social structure of Germanic peoples has been and is the subject of numerous pieces of research; For a review of the problem matter and the relevant literature see K. Modzelewski, *Barbarzyńska...*, Chapter IV, particularly, pp. 206–213.

present this phenomenon we shall make recourse to concrete examples of this type of difference. According to *Edictum Rothari* in the case of a wounding to the face, a freeman was to obtain 16 *solidi*, a half-free man (*aldius*) – 2, while a slave – 1 *solidi*. In the legal code of the Bavarians there was a regulation determining that in a situation whereby the perpetrator deprives the victim of an eye, a freeman was to receive 40 *solidi*, a freed man – 10 while a slave 5, while a freewoman – 80 *solidi*. In the Frisian legal code the regulations on cutting off an ear have the following compensation rates: a nobleman – 16, common freeman – 12, while a half-free (*litus*) – 6 *solidi*. According to *Lex Ribuaria*, in the case of cutting off a hand a freeman was to obtain 100 *solidi* while a slave 18. In turn, in *Lex Saxonum* we find regulations that talk of the compensation for severing the thumb of a nobleman (*adaling*) 360, for a *lit* 30, while for a free girl – 720 *solidi*. These are merely the first randomly selected examples of this type of regulation, a huge number of which are contained in the barbarian *leges*.

Here we are dealing with a complex arrangement of relations amongst three elements: 1) damaged body parts, 2) the amounts of compensation ascribed to them as well as 3) the membership of the victim to various socio-legal categories. The *leges* of barbarian peoples contained greater or lesser developed systems of compensation for damage to particular body parts or 'bodily objects', covering groups of people of varied social-legal status. In individual codes there are various social ranges covered, about which we will talk later. Sets of body parts for the representatives of various status groups are covered within the framework of the said system. Each of them was ascribed a specific sum of compensation in accordance with the victim's status. This was, consequently, its own kind of classification as well as a social valuation of the human body.

Earlier considerations led to the conclusion that the value of compensation reflected the estimated damage of the deed on the basis of the scale of bodily loss. At present, although this conclusion remains in force, the said payments appear within a new context, namely – the differentiation of the legal status of crime victims.

It equally follows to add that in particular *leges* the differences between the amounts of compensation varied in scale. We are dealing with different proportions of compensation amounts determined for subsequent status groups in the case of the same crime against the body. Each of the codes is characterised here by its own course of action. In *Lex Saxonum*, where damages are defined for a nobleman and *litus*, the relationship between their values was 12 to 1, hence, for example, in the case of blood wounding, an adaling could receive 120, while a *litus* 10 *solidi*. In *Lex Thuringorum* there was noted the value of compensation for the nobility and common freemen. The relationship between them was 3 to 1 (for example, for striking an *adaling* – 30, a freeman – 10 *solidi*). Besides in certain legal codes there appear several different proportions depending on the type of crime. An example is here *Edictum Rothari* in which damages were determined for three status groups: freemen, half-freemen (*aldii*) and slaves. There appeared as many as 12 variants in the proportions of compensation amounts.[352] There was a similar situation in *Lex Baiwariorum*, where there existed 7 variant possibilities.[353]

There arises here a fundamental question: how is one to understand the said differences in numerical amounts and the proportions between them? An interpretation of the meaning of the magnitude of these differences and even more their variability is no easy task. It seems possible to accept the view that these proportions were meant to signify a greater or lesser socio-legal distance between the particular categories of victim. If one was to accept such an interpretation then one could claim that in the light of subsequently cited codes for the case of each people the scale of differences of this type was disparate. It is, however, difficult to say how much these proportions reflected actual differences, and how much the convictions of those who compiled the researched regulations in the proper or desired state of affairs in this area. Besides, as one can see on the example of *Edictum Rothari* and *Lex Baiwariorum*, the scales of differentiation depended not only on the status of the victim but also on the type of crime.

In discussing the problem of social differentiation in the values of compensation paid, it follows to draw attention to the distinctiveness between the individual legal codes in terms of number and the type of social groups appearing there. In certain of them there is almost exclusively talk about crimes against the body of freemen[354], while in others in turn against the a noblemen (*adalingi*), ordinary freemen, half-freeman (*liti, aldii*) or slaves, and also against freewomen, dependents and slaves. These are, so they seem, features resulting not so much from (and not only) the actual social situation of particular peoples but

---

352 Most often are applied two variants: a) 6 to 2 to 1 and b) 8 to 2 to 1, besides there appear also: c) 3 to 1 to 1, d) 6 2/3 to 3 1/3 to 1, e) 8 to 3 to 1, f) 8 to 3 to 2, g) 6 2/3 to 3 to 1, h) 4 to 2 to 1, i) 5 to 2 to 1, j) 4 1/2 to 1 1/2 to 1, k) 6 to 2 to 1 1/3, l) 5 1/3 to 2 to 1.

353 The following proportions appear there: a) 4 to 2 to 1, b) 8 to 2 to 1, c) 6 to 2 to 1 1/2, d) 4 ½ to 1 1/2 to 1, e) 10 to 2 to 1, f) 3 to 1 1/2 to 1.

354 In the law of the Salic Franks the amounts of damages were given only for freemen, while in the Kentish code of Æthelberht there existed one single rate for freemen and women. Cf. S. Rubin, *The Bot...*, pp. 147 ff., the author, contrary to the views of other researchers, claims that despite the differences in the values of *wergeld* (for the killing: of a nobleman (*eorl, theng*) – 300 shillings, a freeman (*ceorl*) – 100 shillings) the compensatory rates for damage to particular parts of the body for all social groups was to be the same; his argument concerns title 74 of the Laws of Æthelberht (*Die Gesetze der Angelsachsen*, p. 7), where there is stated that the same rates of compensation which are ascribed to free men should be paid to unmarried women.

from certain assumptions about the ideological character of the legal regulations adopted by the creators of the law. This is a hypothesis whose authenticity we shall consider in a later part of this chapter.

The Reasons for Social Differentiation in the Values of Compensation for Crimes against the Body

A fundamental question connected with the problem of the differentiation in the values of damages as a reflection of the victim's social status concerns the causes of this phenomenon. One may point to its hypothetical sources. The first assumption says that this differentiation resulted from actual, physical (anthropological) differences between the various social groups. This explanation is based on the assumption that the values written into the compensatory tariffs reflected objective, 'qualitative' bodily features of the representatives of these groups. The 'body quality' of one group would be, according to this conception, valued higher, others lower, which would have found expression in the differences in the damage values to be paid.[355]

Such an understanding is not devoid of certain bases. One may here refer to the following arguments: on the basis of research into bone material archaeologists have claimed that women were statistically shorter and smaller than men.[356] This would indicate the existence of actual bodily differences. This in

---

355 The category of 'body quality' was known in the early Middle Ages, for example Liutprand of Cremona (mid 10th century) wrote in *Antapodosis* about the Scandinavians arriving in Byzantium: '*gens* [...] *quam a qualitate corporis Graeci vocant Rusios*' (*Liudprandi Antapodosis*, V, 15, [in:] *Fontes ad historiam aevi Saxonici illustrandam*, ed. A. Bauer, R. Rau, Darmastad 1977, p. 460).

356 Cf. *Burial Archeology: Current Research, Methods, and Developments*, ed. C. A. Roberts, F. Lee, J. Bintliff, 1989: J. Henderson, *Pagan Saxon cemeteries: a study of the problems of sexing by grave goods and bones*, pp. 77–83; *Invisible People and Process: Writing Gender and Childhood into European Archeology*, ed. J. Moore, E. Scott, 1997; S. J. Lucy, *Housewives, warriors and slaves? Sex and gender in Anglo-Saxon burials*, pp. 150–168; B. Effros, *Skeletal sex and gender in Merovingian mortuary archaeology*, Antiquity, vol. 74 (2000), pp. 632–639; R. Hanh, *Die menschlichen Skelettreste aus den Gräberfelder von Neresheim und Kösingen, Ostalbkreis*, [in:] *Die alemannische Gräberfelder von Neresheim und Kösingen, Ostalbkreis*, red. M. Knaut, Stuttgard 1993, pp. 357–428; H. Helmuth, *Antropologische Untersuchungen zu der Skeletten von Altenerding*, [in:] *Das Reihengräberfelder von Alterding in Oberbayern II. Anthropologie, Damaszierung und Textilfunde*, Mainz 1996, pp. 1–143; M. Rouche, *op. cit.*, pp. 440–441, states that amongst the peasant populace (Normandy) the average height of a man was 167 and a for women 155 cm; L. Buchet, *La nécropole gallo-romain et mérovingi-*

turn would influence the value of the *wergelds* and damages designated for men and women. There also exist certain factors coming from archaeological excavations pointing to the fact that the representatives of tribal aristocracies were statistically taller and better built than the rest of the population.[357]

The above argumentation is based on the assumption that the said 'body quality' was the main, if not the only, factor that determined the rate of damages and *wergeld*. This has fundamental weaknesses, which incline one to question the validity of the hypothesis. Firstly, the data contained in the *leges* regulations questions the premise that the value of compensation reflected the actual physical differences amongst the individual categories of crime victims. And namely – in certain legal codes damages and *wergeld* as allocated for free women was higher (usually double) that for men. This example indicates that those who wrote down the law were, in this case, directed by a different criterion. It was possible therefore that in relation to other social groups they determined compensation and *wergeld* on the basis of different factors.

We may examine, however, the connection between the social differentiation of individuals and their corporeal features not simply in relation to the bodily-social realities but also to the sphere of convictions and ideas. There exist certain factors that indicate that amongst barbarian peoples, and at least for the northern Germanic ones, the conviction as to the connection between social position and the physical features of individuals was deeply rooted. The work within which it is expressed is the *Lay of Rig*, which belongs to the mythological cycle of *Edda*.[358] Although it was written down only in the 13th century it seems

---

enne de Frénouville (Calvados), *Étude anthropologique*, Archéologie médiévale, 1978, pp. 5–53.
357 H. W. Bohme, *Adelsgräber im Frankreich: Archäologische Zeugnisse zur Herausbildung einer Herrenschicht unter der Merowingischen Königen*, Jahrbuch des Römisch-Germanischen Zentralmuseum Mainz, vol. 40 (1993), pp. 397–534; cf. also R. Fleming, *op. cit.*, p. 29 ff.; the author describes there the case of an Anglo-Saxon woman of the 7th century of a high social position (a rich grave); she was tall, which was connected with the good healthy living conditions she had enjoyed in childhood.
358 *Rígsþula*, [in:] *Edda. Die Lieder des Codex Regius nebst vervandten Denkmäler*, ed. G. Neckel, vol. I. *Text*, umgearb. Aufl. von H. Kuhn, Heidelberg 1962, lines 2–43; translation of the *Lay of Rig* into English is based on its Polish translation: *Pieśń o Rigu*, [in:] *Edda poetycka*, transl. A. Załuska-Strömberg, Wrocław 1986, pp. 146–152; for the basic literature on *Edda*, cf. J. de Vries, *Altnordische Literaturgeschichte*, vols. I–II, Berlin–Leipzig 1964–1967; G. Turville-Petre, *Origins of Icelandic Literature*, London 1953; S. Einarsson, *History of Icelandic Literature*, New York 1957; T. D. Hill, *Rigsþula: some medieval Christian analogies*, Speculum, t. 61 (1986), s. 79–89; R. Šimek, H. Pálsson, *Lexikon der Altnordische Literatur*, Stuttgard 1987; *Old Icelandic Literature*

that this idea, and maybe also the oral version of the work (myth) which presented its contents, is older and even has a completely archaic character.[359]

The heroes of the *Lay* are Thrael, Karl and Jarl as well as their clans – parents and offspring.[360] In this work it is said that 'a line of slaves came' from Thrael's children[361], while from the offspring of Karl the free heirs descended.[362] There is no direct reference to Jarl's children, that it was from them that the nobilty arose, but his son, Konr, is the ruler.[363] From amongst the various features the work ascribes to these three families there appear also their precise bodily characteristics. Thrael, his wife and offspring are in general ugly – unshapely, strapping, possessing a dark (tawny) complexion.[364] Karl and his rela-

---

and Society, ed. M. Clunies Ross, Cambridge 2000; *The Poetic Edda: Essays on Norse Mythology*, red. P. Acher, C. Larringhton, NewYork 2002.

359 The fundamental and oldest preserved manuscript containing the poetic *Edda – Codex Regius*, is from the end of the 13th century; besides, mythical songs and Nordic mythology have been included in a textbook on poetics by Snori Sturluson, known as *Edda Snorra Sturlusonar*, compiled around 1220 (the oldest manuscript in *Codex Upsalenisis*). It is difficult to determine when the oral versions of the particular parts of *Edda* came into existence as well as being varied; these works were presumably created in the period from the 9th to the 13th century. The *Song of Rig* is preserved in the *Codex Wormianus* (second half of the 14th century), which contains *Snori's Edda*. There is evidence that it was created not in Iceland but in Denmark, for as the ideological source of the work is pagan it seems likely that it came into being after the adoption of Christianity by Denmark i.e., after 965. Presumably it is from the 9th or first half of the 10th century cf. A. Załuska-Strömberg, *Wstęp*, [in:] *Edda poetycka*, pp. XVIII–XXIV i XXXVIII–XLI as well as K. Schier, *Einleitung*, [in:] *Die Edda. Götterdichtung, spruchweisheit und Heldengesänge der Germanen*, transl. F. Genzmer, Köln 1981, pp. 10–14.

360 Thrael's parents are Aï and Edda, Thrael himself had 12 sons and 9 daughters (*Rígsþula*, 2, 7, 12–13); Karl was the son of Afi and Amma, and had 22 children (*Rígsþula*, 16, 21, 24–25); Jarl's parents are Fadir and Modir, Jarl had 22 sons (*Rígsþula*, 27, 34, 41).

361 *Rígsþula*, 13: [...] *þaðan eru komnar þréla éttir.*; the word *þrél* meant 'slave'. F. Genzmer in his translation of *Edda* rendered the Old Icelandic word *Þrél*, which appears there as a first name, with the German word *Knecht* (*Die Edda*, p. 97).

362 *Rígsþula*, 25: *þaðan eru komnar karla éttir*; F. Genzmer (*Die Edda*, p. 99) translated the sentence as follows: *Von diesen stammt der Stand der Freien* (from these came the state of [the] free [people]).

363 Jarl's youngest son is called Konr (*Rígsþula*, 41: [...] *Konr var hinn yngzti.*). The word *konr* meant 'ruler'; *konr ungr* (43) is 'young prince'; Genzmer translated this name simply by means of the word *König* (king), *Die Edda*, p. 102.

364 About Thrael it is said: *Rígsþula*, 7: *Ioð òl Edda* [...] *horfi svartan* [...], which means 'Edda gave birth to a boy [...] dark [as] mud [...]' and later: 8: [...] *var þar a hondum/ hrokkit skinn,/ kropnir knuar,* [...] *fingr digrir,/ fúulligt andlit,/ lotr hryggr, langir hélar*, i.e.,: 'the child had wrinkled skin on its hands /deformed ankles [...]/fat fingers, an ugly face, /a twisted back, long heels" (*Pieśń*, p. 147); his wife, Thír, was gifted with

tions are sturdy folk, robust and ruddy faced; they are not graced by subtle appearance yet do not possess any 'defects' as such.[365] Finally Jarl and his family differed in their bodily nobility and beauty.[366] The said features of appearance, in a similar way to other characteristics (intelligence, wealth, life style, occupations) for the heroes were as if 'inborn', or at least fully developed by the time Rig - i.e. Heimdall (Odin's son) wandered round the earth in human form – visited the future parents of Thrael, Karl and Jarl. The social status of the members of each of these three families was, according to the account in the *Lay*, defined by Rig somehow on the basis of their 'primordial' situation. In this way there was established and at the same time sanctioned the social order. It results from this that the three heroes, and even more so their parents, were not initially assigned to a definite social group. At this stage of the history of the community there still did not exist a formalized division into social or status groups. It follows, however, to note that already the parents of our heroes were equiped with traits of status.

The *Edda* account supposed the existence of a link between the affiliation of an individual to one of the three social groups and its corresponding bodily features. It is difficult to evaluate whether the creators of barbarian laws shared this conviction. The *Lay of Rig* presented all the same an initial, model situation. It follows, however, to clearly state that the vision of society contained in it cannot be treated as an explanation as to the genesis of legal order. The three social groups mentioned therein were not legally categorised. The idea of uniting particular social 'orders' possessing certain bodily features as equally a specific 'body quality' could, however, have been known to the creators of the codes of barbarian laws and constitute for them an ideological point of reference. It is not excluded that those compiling the barbarian *leges* based their reasoning and actions on the following postulate: the social status of an individual is connected to his/her bodily features. It does not result, however, from this principle that the said status and bodily features jointly defining it translated itself into some form of definite value, and particularly a material value measured in money or kind. On the other hand, the view that the 'body quality' of man constituted an element in his legal position, would be conducive to the attributing of a definite

---

similar features – she was swarthy, with a concaved nose (*ibidem*); *Rígsþula*, 10: [...] *armr solbrunnin/ niðrbiuget er nef* [...].

365 About Karl: *Ióð òl Amma/* [...] *rauðan ok rioðan/ riðuðu augu.* (*Rígsþula*, 21), which means 'Amma gave birth to a boy [...] red-headed and ruddy faced, he moved his eyes'.

366 About Jarl it is said: *bleikt var har/ biartir vangar/ otul voru augu/ sem yrmlingi* (*Rígsþula*, 34), which means: 'flaxen was [his] hair/ clear cheeks, sparkling were [his] eyes (piercing gaze?)/ like a small snake has'.

material value to particular parts of the body. This was an initial condition, though an insufficient one, for the creation of such a link.

The differences in the values of compensation – as we have established – did not result directly from distinctiveness of the bodily features of particular social groups. We are deliberating, consequently, over a different hypothesis to explain this phenomenon. The distinctiveness in this field may have resulted from the social arrangement of forces as manifested amongst the individual groups. A nobleman whose hand was cut off *did not lose* 'objectively' any more (i.e., physically) than an ordinary freeman even though the *welgeld* for the former was higher than for the latter. The reason for the difference may have been that the magnates demanded greater rates of compensation than ordinary freemen. For as they possessed after all realistically more effective means to execute their dues than was the case for other social groups such demands were satisfied. On the other hand, in a case whereby the magnate was the perpetrator of the crime he was equally more able than others to pay even a high rate of compensation.[367] In this way social differences would have influenced the rate of compensation for various violations of the body. Such an explanation referring to the arrangement of forces within group relationships, although reasonable, appears insufficient. Possibly the compensation amounts awarded to particular groups was suppose to be its own form of sign as to the strength and authority that was possessed by each of them in the social distribution of force. This thesis has equally a limitation. For it refers only to freemen. It does not explain, therefore, the situation of freewomen who were also at times given higher rates of *wergeld* and compensation than men. Therefore this is one of a few explanations.

It results from the above considerations that the search for the causes of the said differentiation within the sphere of 'objective circumstances' (bodily features, the social distribution of strength or others still) leads to seeming rationalisations, while at the same time providing only a partial answer that covers certain social groups to the exclusion of others. This conclusion invites the assumption that a significant source of the discussed phenomenon was the ideological

---

367 Cf. K. Fischer Drew, *The family in Frankish Law*, [in:] eadem, *Law and Society in Early Medieval Europe*, London 1988, VI, pp. 3–4; the author, in commenting on the regulations of Salic Law (*PLSal*,XLII, 1–3 and XLIII, 3), writes about a magnate who commanded armed bands (*contubernia*); the commander of such a group paid an increased level of compensation (600 *solidi*), for he killed a free Frank, having carried out an intrusion on his property together with his band; other members of the group paid 90 or 45 *solidi*. Drew claims that the murder carried out by the said group was treated differently than if conducted by a single individual. Relatives obtained in the first case a higher rate of compensation on the condition that they were able to bring the band to court.

convictions or assumptions of the creators of the *leges*, who in a 'model' way defined the situation of the particular groups of people.

Tribal society was composed of many categories of people with various legal statuses and positions. These groups created a hierarchical structure. The Germanic codifiers were convinced that freemen (ordinary and noble), freed men and slaves were not and could not be equal. Therefore, in the compensation tariffs and rates each of these categories was examined separately. Inequality in status was the basic feature and principle in the functioning of these societies. It follows to suppose that for those arranging the law (in a similar way for their compatriots) the fundamental difference amongst the mentioned groups was so obvious that they could not imagine society without it. K. Modzelewski, in considering the question of the differences amongst the particular barbarian social categories (particularly the Saxons and Frisians), has advanced the thesis that they were derived from a determined 'quality of person' resulting from birth. This concept, in the author's opinion, had equally a moral character.[368]

The inherent superiority or inferiority of particular categories of people of noblemen (*edelings*) and commoners (*frilings*) was expressed in the various rates of *wergeld* and damages attributed to them. This principle equally applied itself to half-freemen (*liti*) and slaves. J. Smith, who links *wergeld* and compensation with the regaining on the part of the victim or his family of the honour of the individual and clan, tarnished by murder or wounding, claims that the gradation of these rates in relation to the ranking of the victim shows that individuals with a high social position risked losing it to a greater degree than for those situated lower down the hierarchy.[369] On the one hand, their personal and family

---

368 K. Modzelewski, *Barbarzyńska*..., p. 246; the author refers to the regulation of *Lex Saxonum* (Title XVII), in which it was decided that a *friling* (common freeman) was to be cleared of the charge of the murder of someone else's slave together with 12 joint oath takers, while the *edeling* (nobleman) in this same case was to appear with 4 oath takers; the *edeling*'s oath had greater power as a result of his noble birth; cf. also *Lex Frisionum*, titles I and II; Frisian *nobilis* in the case of a charge of killing another noble he was cleared together with eleven individuals, while if the matter concerned the murder of a *litus* three people were sufficient; however if a *litus* was charged with murdering a nobleman he had to swear with 36 people; if he killed another *litus* – with 12, if a freeman – with 24.

369 J. Smith, *op. cit.*, p. 101; the author in her exposition on the level of compensation rates refers to the regulation of Bavarian law, in which there is mention of five nobly born families who were entitled to double the *wergeld* rate than for ordinary freemen as a result of their 'double honour'; Cf. *LBaiw*, III. *De genealogiis et eorum compositione*, 1. *De genealogia, qui vocatur huosi, drozza, fagana, hahiligga, anniona: isti sunt quasi primi post agilolvingas, qui sunt de genere ducali; illis enim duplam honorem concedamus, et sic duplam conpositionem accipiant.*

honour was worth more than amongst ordinary freemen or half-freemen (*liti*), while on the other, as a result of fulfilling leadership functions they were more subject to any disgrace to their reputation brought about by wounding.

The inequality of the positions of particular 'orders' within the tribal collective manifested itself in a number of ways. Of particular interest given our subject matter, although by no means the only one, in indication of status are the values of compensation for violating the body of representatives of particular groups. We shall state clearly, there is nothing obvious in the fact that the 'valuation' of the rates of compensation for bodily violation depended on which category of people a given victim belonged. This was the codifiers' premise that directed them in the drawing up of compensatory rates. The existence and effect of this regulation is by no means obvious on its own. The principle arose and was applied in societies of a specific type (at a definite stage of their development) and at a definite historical epoch. Outside of these cultural and temporal frameworks this principle lost its obviousness and stopped being in force.[370]

Summing up, the rates of compensation presented in the *leges* in the case of individual crimes against the body depended on the legal-social status of the victim, that is on the belonging of a said individual to one of the groups defined by those codifying the law. Such factors as: the potential for means of force resulting from a family's strength, authority over people and the material wealth of an individual; the actual differences in bodily features, the convictions concerning the relationship between social position and 'body quality', to various degrees conditioned the social and legal position of an individual. It did not, however, determine the direct rate of compensation and did not result in its differentiation. They influenced it merely indirectly through the category of legal status and the rules regulating the relations between groups.

There remains, however, the question as to whether the said legal status had in the light of the researched *leges* some reference to the bodily sphere. There exist accounts that such a connection existed in the case of freemen. For in *Edictum Rothari* there is talk that the *wergeld* of a freewoman (in a similar way to men) depended on the nobility of her birth (*generositas, nobilitas*).[371] In the case of certain crimes the value of a woman's blood money could have been same as

---

370 In the codifications of Roman Law functioning in parallel with the barbarian codes, i.e., in *Breviarium Alaricii* or *Lex Romana Burgundionum*, which constituted a simplified version of Roman law from the imperial period (particularly *Codex Theodosianus*) there was no principle of paying compensation for crimes against the body, and with the same there was no social differentiation between the said payments.

371 *ERot*, 75. *De infante si in utero matris occisus fuerit.* [...] *si ipsa mulier libera est et evaserit, adpraetietur ut libera secundum nobilitatem suam.* [...] *Nam si mortua fuerit, conponat eam secundum generositatem suam* [...].

the payment for the killing of a man of 'the same blood' (*de similem sanguinem*), and specifically – of her brother, or some close relation.[372] The fines and *wergeld* of freeman (and women) were defined in relation to the 'quality of the person' (*adpraetietur, qualiter angargathungi, id est secundum qualitatem personae*).[373] 'Nobleness' or '(high) birth', 'quality of person' and 'blood' were concepts closely linked to each other, mutually resulting from each other. Legal status was therefore defined through the bond of kinship, and correspondingly its connection with the bodily or biological aspect of a man's life. In this sense the legal status of freemen was connected with the corporeal sphere.

Images of Germanic Social Structure and the View of the Human Body as Perceived by the *Leges* Codifiers

One of the characteristic features of the Germanic *leges* is the specific link between the classification of social groups within the area of each of these peoples and the systems of regulations on crimes against the body. Before we present the nature of the said relationship, we shall discuss briefly the structural vision of certain Germanic peoples contained within their legal codes. It should be emphasised that the aim of our investigation is not a reconstruction of the actual social character of the tribal collectives herein described.[374] This results in part from the nature

---

372 *Ibidem*, 187. *De violentias mulieris libere.* [...] *Nam si contegerit casus, ut,* [...] *mortua fuerit* [...] *ille, vir, qui eam violento ordine tulerit uxorem, conponat eam mortua; tamquam si virorum de similem sanguinem, id est fratrem eius occidisset, ita adpretietur* [...].

373 *Ibidem*, 48. *De oculo evulso*; cf. also, *ibidem*, 11. *De consilio mortis.* [...] *et si ex ipso consilio mortuus fuerit, tunc ille, qui homicida est, conponat ipsum mortuum, sicut adpraetiatus fuerit, id est wergild.*

374 There exists extensive literature on the subject: K. Bösl, *Frühformen der Gesellschaft im Mittelalterlichen Europa: Ausgewälte Beiträge derMittelalterlichen Welt*, Münich 1964; D. A. Bullough, *Early Medieval Social Groupings: The Terminology of Kinship*, Past and Present, 45 (1969), pp. 3–18; A. C. Murray, *Germanic Kinship Structure. Studies in Law and Society in Antiquity and the Early Middle Ages*, Toronto–Ontario 1983; D. Whitelock, *The Beginnings of English Society*, Harmondsworth 1972; K. Fischer Drew, *Introduction*, [in:] *The Lombard Laws*, transl. K. Fischer Drew, Philadelphia 1973, pp. 6–37; Ch. Wickham, *Early Medieval Italy: Central Power and Local Society*, London 1981; R. Schmidt-Wiegand, *Fränkische und frankolateinische Bezeichnungen für soziale Schichten und Gruppen in der Lex Salica*, Nachrichten der Akad. der Wissenschaften im Göttingen I, Phil.-hist. Klasse nr 4, Göttingen 1972, pp. 219–253; K. Fischer Drew, *Introduction*, [in:] *The Laws of the Salian Franks*, pp. 39–49; I. Wood, *The Merovingian Kingdoms 450–751*, London–New York 1994; P. D. King, *Law and*

of the legal regulations themselves. As their function was not to comment in a detailed and exhaustive way on the specifics of the social structure.

The fundamental aim of law was to determine the legal position of particular individuals and their groups, as well as the regulating of the relations amongst them. This feature of the researched legal texts had a large influence on the relations of presenting various, otherwise real, social phenomena. The codifiers editing the individual regulations (out of necessity) utilised certain thought schemes useful in the creation of model conceptions of society, which did not have to faithfully reflect reality. They reflected well, however the requirements of the regulatory function of the law. We shall not claim that the listings of Germanic laws do not contain information on their actual social structure. The matter rather concerns the fact that knowledge of this type was used by particular barbarian codifiers to create classifications of social groups which corresponded to their vision of their people. It follows to here add that the presentation of the said vision was not the main aim of the codification. The authors of the laws of particular Germanic codes had to, at least for their own use, formulate a model of the structure of their own people reflecting their priorities and experiences.

'Images' of the Social Structure of Germanic Peoples – a Comparative Analysis of Selected Examples.

The classifications of social groups contained in the laws of individual Germanic peoples may be reconstucted on the basis of regulations that talk of killings and bodily violations. We shall begin our look at different variants of the said classifications from a discussion of *Lex Frisionum*.[375] Title I (*De homicidiis*) of this code contains regulations on the killings of representatives of various social groupings. Here are enumerated four groups of people: noblemen (*nobiles*), freemen (*liberi*), freed men (*liti*) and slaves (*servi*). The paragraphs from 1 to 4 talk of killings committed by *homo nobilis* subsequently on: other *nobilis*, freemen and those freed. Paragraphs 5, 6 and 7 concern killings carried out by a free man on: *nobilis*, other freemen and those freed. Points 8, 9 and 10 regulate cases

---

Society in the Visigothic Kingdom, London 1972; C. Hammer, *A Large-Scale Slave Society of the Early Middle Ages: Slaves and Their Families in Early-Medieval Bavaria*, Aldershot 2002; K. Fischer Drew, *The Law of the Family in the Germanic Barbarian Kingdoms: A Synthesis*, [in:] eadem, *Law and Society in Early Medieval Europe*, London 1988, VIII, pp. 27–36; G. von Olberg, *Die Bezeichnungen für soziale Stände, Schichten und Gruppen in der Leges Barbarorum*, Berlin–New York 1991; K. Modzelewski, *Barbarzyńska...*

375 Cf. H. Siems, *Lex Frisionum*, [in:] *HRG*, vol. II, col. 1916–1922; idem, *Studien...*, pp. 222–232.

of a killing by a freed man of *nobilis*, a freeman and another freed man. Point 11 concerns the killing of someone else's slave by 'some man' (*quis homo*), i.e.,: *nobilis*, a freeman, a freed man or another slave. Paragraph 13 regulated the case of the murder of a *nobilis*, a freeman, or freed man by a slave without his master's knowledge. Paragraph 14 to 18 speak of the killing of a *nobilis*, freeman or freed man committed by a slave on the order of his master.

The *wergeld* for killing a man defined as 'nobilis' (*adaling*, noblemen) was 80 *solidi*.[376] This amount was not dependent on the perpetrator's status. The same principle applied in the case of freemen and slaves. The *wergeld* of a freeman was 53 *solidi* and 1 *denar*[377], while the blood money for a freed man (*litus*) was 26 *solidi* and 2 *denari*. In the case of a *litus* the amount was given to his master, while the family of a murdered man was to be paid by the perpetrator somewhat less than 9 *solidi* (*solidos IX excepta teria parte unius denarii*).[378] Slaves were not allocated compensatory amounts. Paragraph 11 says that the perpetrator *componat eum iuxta quod fuerit adpretiatus*, and therefore according to an individually established value for a given slave.

Attention is drawn by the proportions between the actual sums represented by the respective *wergelds*: 80 *solidi*, 53 *solidi* and 1 *denar* as well as 27 *solidi* minus 1 *denar*, that is respectively 100, 66 and 33%.[379] It follows to add that the sum of 9 *solidi* without one third of a denar, which was received by the family of a murdered freed man, was equally written into this scheme. For it constituted 1/3 of the sum designated for an ordinary freeman (27 sol. minus 1 den.), i.e., 11% of a noble's *wergeld*. Therefore we are dealing with a systematic solution, schematic in character.

This is directed by the assumption that the said three groups (the nobles, freemen and freed men) were internally uniform. This system introduced, or at least implied, a well advanced standardisation of the three 'orders'. The rigidly

---

376 *LFris*, I, 1. *Si nobilis nobilem occiderit, LXXX solidos componat* [...]; 5. *Si liber nobilem occiderit, LXXX solidos componat* [...]; 8. *Si litus nobilem occiderit, similiter LXXX solidos componat* [...].

377 *Ibidem*, 3. *Si nobilis liberum occiderit, solidos LIII et unum denarium solvat* [...]; 6. *Si [liber] liberum occiderit solidos LIII et unum denarium solvat* [...]; 9. *Si [litus] liberum occiderit, solidos LIII et unum denarium solvat* [...].

378 *Ibidem*, 4. *Si nobilis litum occiderit* [...]; 7. *Si [liber] litum occiderit* [...] i 10. *Si [litus] litum occiderit* [...].

379 These *wergeld* rates were in force in middle Frisia ; for wealthy freemen and freed men from western Frisia different values were designated – respectively 100, 50, and 25 *solidi* (*LFris*, I, 10), while different again for eastern Frisia– 106 sol. and 2 den., 53 sol. and 1 den. as well as 26 1/2 sol. and 1/2 tremisse (*ibidem*); cf. H. Siems, *Studien*..., p. 223; idem, *Lex Frisionum*, [in:] *HRG*, vol. II, col. 1920.

defined amounts for blood money of the said three groups contrasts with the regulation on compensation for the death of a slave, which was individually defined.

The amount of blood money was therefore accredited to the given status grouping in a permanent way. Attention is also drawn by the fact that in the discussed title of Frisian law there is no talk whatsoever about women. While it is known that title XI, *De farlegani* (On Fornication), was devoted to women of various status. Here are mentioned 'noble' women, freewomen, freed women and slaves.[380] There is, however, an absence in the codification or children or juveniles.

In *Lex Frisionum* we are therefore dealing with a model representation of tribal social structure. Its dominant was the hierarchical arrangement of three groups of men: the *nobiles*, freemen and the freedmen, supplemented by the category of slaves. This was a consciously created structure for the needs of legal regulation, presenting the social situation of the Frisian community in the form of a scheme. It is a suggestive vision thanks to its simplicity. It does not seem that it could have fully (as a result of its character) reflected the richness of social reality.

We can observe a similar vision of the arrangement of social groups to that of compensation in title XXII *De dolg* (On Wounds). It contains 87 paragraphs on various aspects of violation to varied parts of the male body as well as the amounts of compensation allocated for them. From our point of view the most important is *Epilogus*, located at the end of the title. For this illuminates how its contents should be understood. Firstly it is here stated that the compensation enumerated in paragraphs 1-89 concerns freemen. Secondly, that the discretionary damages *homo nobilis* are one third higher than those for freemen, and thirdly, that in the case of *liti* (freed men) the compensation is a half of the amount for freemen.[381] Women are also not taken into consideration here with the exception of the last two points (88 and 89) that talk of grabbing a freewoman, somebody else's woman by the breast and vulva.

It is equally important to draw attention to the proportions between the amount of damages for the three groups mentioned and the way the principles for their designation was presented. In the case of the *nobilis* they were one third higher than for freemen, while freed men (*litus*) were allocated compensation at

---

380 LFris, IX, 1 – *femina nobilis et libera*, 2 – *lita*, 3 – *ancilla*, 8 – *puella virgo, nobilis sive libera*, 10 – *puella lita*.

381 LFris, XXII, (*Epilogus*) *Heac omnia ad liberum hominem pertinent. Nobilis vero hominis compositio, sive in vulneribus et percussionibus et in omnibus que superius scripta sunt, tertia parte maior efficitur. Liti vero compositio, sive in vulneribus sive in percussionibus sive in mancationibus et in omnibus superius descriptis, medietate minor est quam liberi hominis.*

a rate half that of freemen. In the case of compensation for murder the matter is presented as follows: a free man had a *wergeld* double the rate of a freed man (*litus*) (53 sol. and 1 den. and 27 sol. minus 1 den.); in other words the damages for a *litus* were a half of the blood money for an ordinary freeman. While the *wergeld* of the *nobilis* was a half (50%) higher than for a freeman (80 sol. and 53 sol. and 1 den.); this means that the blood money of a freeman constituted two thirds of that for a *nobilis*. The proportion amongst the rates of damages was, therefore, somewhat different than in the case of blood money (100, 66 and 33%). They were expressed by the following values 133, 100 and 50% (or also: 100, 75, 37.5%, if we are to take as the starting value the nobleman's damages).

The question of proportions is not reduced, however, to percentage divisions. For we can note that in *Epilogus* title XXII the value of the damages for *nobilus* and freemen was designated on the basis of the damages of a freeman. This was adopted as being 100%. This suggests that the compensation of freeman was primary in relation to the remainder. One may suppose that initially a list of damages for a freeman was drawn up and later there was specified the principle upon which the compensation for the *nobiles* and freed men was calculated. This is an important conclusion for us to draw. For this principle shows the role that the Frisian codifiers attributed to free people, and strictly to freemen. It seems that according to their convictions freemen were the core, its own form of centre of gravity for the community, and simultaneously the model through which it was possible to define the position of the remaining 'orders'.

In this sense the image of the structure of Frisian society differs from that emerging from the title *De homicidiis*. For there the description of the hierarchical arrangement of society began from the group with the highest social status i.e., the nobility. However, it is not to be excluded that in this case the *wergeld* of the nobles and freed men (*liti*) was a derivative of the blood money of freemen; 26 *solidi* and 2 *denar* (*litus*) is a half of the 53 *solidi* and 1 *denar* (freeman), therefore 80 *solidi* (noble) is the sum of these figures, i.e., the *wergeld* amount of a freeman increased by a half.[382] Another hypothesis is also possible, namely that the starting point was here the sum of 80 *solidi*, from which there was initially taken a third of it, and then two thirds which resulted respectively in 53 *solidi* and 1 *denarus* as well as 26 *solidi* and 2 *denari*.

The Frisian system of damages which, as we have stated above, possesses features of a holistic solution, omitted not only slaves but also women. This

---

382 Cf. H. Siems, *Lex Frisionum*, [in:] *HRG*, vol. II, col. 1920, who claims that on the basis of an analysis of title XXII *liberi* constituted in Frisian law 'the main social estate', on the basis of damages for them was calculated the rate of compensation for the nobles and freed men (*litus*).

concerned all the status categories of women: noble women, ordinary freewomen, freed women and slave women. Whereas from title IX of *Lex Frisionum*, concerning sexual crimes against women, we know that the structure of the female population was in its own way copied from the image of the male collectivity known from title I (*De homicidiis*). We should here draw attention to the fact that the said title I did not contain a point that talked of the *wergeld* of women. This matter was regulated in *Additio sapientum*. *Wlemarus*, in title V (*De muliere occisa*). Here there is talk that for the killing of a woman (*mulier*) the perpetrator was to pay blood money *iuxta conditionem suam, similiter sicut et masculum eiusdem conditionis solvere debet*.[383]

It is not clear either how title V of *Additio* is to be understood. It seems that the word *mulier* meant in the clause a woman in general, regardless of her social position. This explanation finds confirmation in a further part of the quote: *wergeld* was to be set 'according to her rank, similarly to a man (*masculus*) of this same condition'. The word *masculus* is here used intentionally. For it means (using a descriptive expression) a male individual.[384] Analogically to the case of women it consequently referred to men in general. Both communities – male and female – were divided according to the criterion of status (*conditio*) into several categories. Both in title I and in title IX of *Lex Frisionum* there occur four 'orders': the nobility, the common freemen, the freed (*litus*) and slaves. Therefore, we can assume that Wlemar, in talking about the differentiation of *wergeld* in relation to social-legal status had in mind this particular four-divisional structure of the Frisian community. In the case of *Additio*'s title V this division related to women.

However, let us return to the question of omitting the fines of slaves. It is difficult to resist the conclusion that we are here dealing with an intentional act on the part of the codifiers. On the other hand, in other rules in the code (titles I and IX) compensation for crimes against life and body covered slave men and slave women. Therefore, it seems that in this sense *Lex Frisionum* displays a significant constructional incoherence. However, this is not an isolated case. In the law of the Thuringians, where there is allocated as compensation for murder: to noblemen (*adalings* - 600 *solidi*), ordinary freemen (200 *solidi*) and slaves (30 *solidi*) as well as to freed individuals (80 *solidi*), there is also an absence of damages for the last two categories as opposed to the nobles and freemen.[385]

---

383 *Add. Sap.*, V, 1, p. 98.
384 In the German translation it is used as the equivalent to the word *Mannperson*, (*L. Fris.*, transl., K. A. Eckhardt, p. 99).
385 *LThur*, I–III i XLIII (murders) and IV–XXV (bodily damage).

What is more there are also absent regulations on crimes against a woman's body.

* * *

Another example of the classification of the social groups is the Lombardian *Edictum Rothari*.[386] In a way different than in Frisian law, in Rothari's codification the vision of social construct was recorded first and foremost in the system of damages. The regulations on murders are in this case less representative although that may equally be incorporated into our research. The rules regulating crimes against the body were arranged in the discussed code according to social 'orders'. The first group of regulations (the titles from 42 to 73) concern freemen. Title 42 has the heading *De homine libero legato*. Title 74, concluding this section of regulations, contains pieces of information that the crimes described in the preceding regulations occur amongst freemen (*quae inter hominis liberos eveniunt*). The second group of regulations (the titles from 77 to 102) concern half-freemen (*aldii*) and individuals termed as *servi menisteriales*. The status of the latter is explained by Title 76 – *menisteriales dicimus, qui docti, domui nutriti aut probati sunt*. Finally, the third group of regulations (the titles from 103 to 126) concern a category of individuals described as *servi rusticani* (slave-peasants). Title 127 contains an important supplement to these regulations. Here is talk that the above regulations refer equally to men as to women and more precisely freed women and slaves (*aldias aut ancillas*). Here it follows to add that in the regulations on freemen there is a lack of information about freewomen. However, there exist sources that state their damages were of the same amount as in the case of men.[387]

There emerges from this a picture of a three-part structure for Lombard society, composed of: 1) freemen, 2) the freedmen and educated slaves as well as 3) rural slaves and commoners, both men and women. It follows to add that the picture of the group of freemen is here somewhat unclear. For on the one hand, it presents them as a single category, while on the other, in several titles the rates of damages were designated in the form of a fraction of the *wergeld*, which in turn was to be determined according to the quality of the person (*in an-*

---

386   On the subject of the social structure of the Lombards in the 7th to 8th century cf. K. Fischer Drew, *The Lombard Laws...*, pp. 28–31; K. Modzelewski, *Barbarzyńska...*, pp. 218–231.

387   *ERot*, 378. *Si mulier libera in scandalum cocurrerit* [...] *si plagam aut feritam factam habuerit* [...] *adpretietur secundum nobilitatem suam et sic conponatur, tamquam, si in fratem ipsiusmulieris perpetratum fuisset*; cf. also Title 187, where there is talk of the self same way of determining compensatory amounts in the case of the death of a raped woman, i.e., according to the rate due to her brother.

gargathungi, id est secundum qualitatem personae).[388] This means that the said 'quality of person' was different amongst freemen, and as a consequence the difference was also the value of the blood money for the particular individuals. However, in *Edictum Rothari* the names of the status categories of freemen are not mentioned nor are any rates for *wergeld* designated for them.[389]

Another means of recreating the picture of Lombard social structure is research into the regulations on killings and the rates of compensation contained in them. A series of regulations determining the rate of compensation for killing people of varied position was included directly after the above discussed regulations on damages for bodily violation. These are the titles from 129 to 136. The following categories of person and ascribed rates of compensation are mentioned: *aldii* – 60 *solidi*, *servi ministeriales* (i.e., qualified household slaves) – 50 *solidi*, other *servi ministeriales* (uneducated) – 25 *solidi*, *servi massari* (slave-lessees) – 20 *solidi*, *servi bovoculi* (slaves ploughing with oxen) – equally 20 *solidi*, *servi rusticani* (slave-peasants) – 16 *solidi*, *pastores-porcarii* (pig herders) – 50 *solidi*, *porcarii* of the lowest ranking – 25 *solidi* as well as *pecorarii*, *caprarii* or *armentarii* (shepherds, goatherds and cattle herders) – 20, and their apprentices (*discepulos*) – 16 *solidi*.[390]

As can be seen the classification of the varied categories of people partially or totally unfree, compiled to designate the rates of compensation for murder, was far more complex than the one resulting from the gradation of damages. For here are listed, beside *aldii*, eight groups of slaves differentiated amongst themselves in terms of education (qualifications) and type of occupation.

---

388   *ERot*, 48. *De oculo evulso*; this same principle was applied towards *aldii* and slaves, although it was expressed in different words, cf. 81. *De oculo evulso. Si quis aldium alienum aut servum ministerialem oculum excusserit, medietatem pretii ipsius, quod adpretiatus fuerit, si eum occidisset, ei conponat*; as well as 105. *De oculo evulso. Si quis servum alienum rusticanum oculum excusserit, medietatem pretii ipsius, quod adpretiatus fuerit, si eum occidissit, dominum eius conponat.*

389   On the subject of wergelds cf. K. Modzelewski, *Barbarzyńska...*, p. 200; the values of the wergelds of free Lombards of various 'quality of person' were entered into Title 62. of *Liutprandi leges* of 724; here there is talk of two categories of freemen-warriors: 'the smallest people' (*minimae personae*) as well as about the 'first' (*primi*), who were according to custom entitled to a *wergeld* of 150 and 300 *solidi* respectively, as well as about two categories of *gasind* – 'the smallest' (*minimi*) and 'larger' (*maiori*), to whom the Liutprand designated wergelds at a rate of 200 and 300 *solidi*.

390   *ERot*, 129–136; cf. the translation into English of the Latin terms for the above categories of slave (K. Fischer Drew, *The Lombard...*, pp. 72–73): *household slave, tenant slave, ox ploughman, herder, cattle herder, goat herder, ox herder*; in the Italian translation (C. Azarra, *Leggi...*, pp. 37 and 39) the words: *bovaro, pecoraio* and *armantario* are used.

In the light of the rates of compensation the differentiation amongst the slaves was somewhat different, for the codifiers designated four rates - 50, 25, 20 and 16 *solidi*. It is worth drawing attention to that fact that in the regulations under discussion there is talk of their gradation, and also the dependence of slaves of one type in relation to others (e.g., slave-peasants, slave-lessees or apprentices in relation to a shepherd).[391] This differentiation was expressed by means of a gradation of the rates compensation for killing – e.g., a qualified household slave 'was worth' 50, while an unqualified – 25 *solidi*.

Edictum Rothari contained also regulations on the murder of free women (married, unmarried and maidens).[392] In all three cases the rate for blood money was the same – 1200 *solidi*, in which the family of the victims received 600 solidi, while the other 600 went to the king. It follows, however, to emphasise that the titles on women's murders were located in another part of the *Edictum*, detached from the above discussed regulations. In this same part there are to be found also regulations on the rape of an *aldia* (half-free woman), freed woman (*liberta*) and a slave (*ancilla*).[393] The compensation was in turn 40, 20 and 20 *solidi*. These regulations constituted a supplement to the image of Lombard social structure contained in the researched code. They contain the category of freed woman (*liberta*), equal in status to a slave, the equivalent to which was absent in the male populace.[394]

As can be seen the reconstructed classification of the arrangement of Lombard social groups was composed of several different elements. The tariffs of fines for crimes against the body, as we have stated, was composed according to a scheme of dividing society into three status categories: freemen, *aldii* and slaves. The group of titles for damages and compensation for murder presents one with a different picture of society. As we have already recalled there was an absence amongst them of a *wergeld* for a freeman. Yet the image of the slave 'order' in the light of these regulations is significantly more complex than within the three-part scheme. In the regulations referring to bodily violations there appear two categories of slave: *servi ministeriales* and *servi rusticani*. In the second group of titles we have beside these two another seven different categories of slave. The attention of the codifiers was concentrated first and foremost on

---

391 *ERot.*, 134. *De servo rusticanum, qui sub masssario est* [...]; 136. *De pecorario, caprario seu armantario, magistro tamen* [...] *pro discepulos autem, qui sequentes sunt* [...].
392 *Ibidem*, 200, 201.
393 *Ibidem*, 205–207.
394 Cf. K. Modzelewski, *Barbarzyńska*..., p. 190, the author comments on titles 205–207 and explains the difference in compensation (40 and 20 *solidi*) between an *aldia* and a liberated woman ; the first was born as a free person, i.e., her mother was a free woman, while a liberated woman was first generation free.

them. However, there is no information here about the compensation paid for the murder of freed women (*aldia*) and slave women. On the one hand, we have therefore an attempt at presenting a picture of the whole of Lombard society (although there is an absence within it of free women) in the form of a simplified scheme. While on the other hand, this is a fragment of a picture in which one status group is presented in an extremely detailed way, one which appears to be closer to reality. However, of the two remaining social groups one (the freemen) has been omitted, while the other (the dependants) treated in a marginal way. For there is no designation as to the compensation to be paid for the killing of a freed man (*libertus*). The regulations on rape show that apart from the half-free woman (*aldii*) there existed also a category known as *liberti*, that is freed people of the first generation.

The parts of the *Edictum* that talk of crimes against the body and about killings may be interpreted as a single 'section'. This would be borne out by their location next to each other in the text as well as the similar nature of the crimes regulated by them. The second of the listed sections would be, according to this interpretation, a development of the first. Such a means of interpreting the researched fragment of the Lombard code has, however, the weakness that in the section that talks about killings there appear only two social groups: the *aldii* and the slaves. Here there is also no mention whatsoever of women. One can see therefore that these two schematic images of the structure of society do not correspond to each other. It seems that at their basis lay dissimilar assumptions and goals. In a collection of regulations containing rates for damages, the codifiers concluded a frame vision of the social order. Yet the titles on killings served, so it seems, chiefly to regulate the situation of slaves. For attention is drawn to that fact that in relation to a freeman and a freed man the regulations of the law were formulated in such a way in order for their internal differentiation not to be accentuated. While in the case of slaves the attention of those encoding the law was concentrated first and foremost on the diversity of this social group. However, it appears that the creation of separate regulations for certain categories of slaves did not have a practical justification; for *servus massarius*, *bovoculus* as well as *pecorarius*, *caprarius* or *armentarius* were evaluated in the same way.

* * *

Let us move in turn to the codification of the Salian Franks. The main material here for a reconstruction of the perceived arrangement of social groups is the regulations on murder.[395] Regulations talking about violations to the body can be helpful only as comparative data. It is important to note straightaway that the

---

395 Cf. K. Fischer Drew, *Introduction*, [in:] *The Laws of Salian Franks*, pp. 45–49.

account provided by particular manuscripts of Frankish legisation is not uniform for the matter in question. It is necessary as a consequence to take into consideration the differences between *Pactus legis Salica* and *Lex Salica*. The common feature of both versions is, however, that in each of them the regulations on the killing of people of various socio-legal status was noted down in several separate titles. In *Pactus legis Salica* these regulations are to be found in five different places, in titles: XLI, XXIV, XLIV, LXVe and XXXV. The reconstruction of a coherent picture of the structure of Frankish society is no easy task. For the regulations talking about such similar crimes have been written down in two and even three of the mentioned titles. However, what is more sometimes they differ with regard to content. It is therefore essential to examine more closely the particular regulations and to compare them with each other.

The largest number of regulations that deal with the matter of interest to us are to be found in title XLI (*De homicidiis ingenuorum*), which contains 21 paragraphs. Blood money was here not arranged according to the value of the rates, but in a different way. The *wergeld* of a freeman, a Frank or some other 'barbarian' (i.e., Germanic man) living in accordance with Salic Law was 200 *solidi* (§ 1). The subsequent three paragraphs concern the hiding of the corpse of a freeman. In point 5, a 600 solidi *wergeld* was set for the killing of a man *in truste dominica*, that is a Frankish courtier[396] as well of a freewoman. Paragraphs 6 and 7 concern the hiding of the body of the said dignitary. Points 8, 9, and 10 regulate the killing of three categories of Romans. In turn the murder of a free girl who had yet not give birth required compensation at a rate of 200 *solidi* (§ 15); of a woman who had started to breastfeed 600 *solidi* (§ 16), while in the case of one who had reached middle age and was unable to give birth – 200 *solidi* (§ 17). The blood money for a long-haired (uncut) boy was 600 *solidi* (§ 18). Similarly, the killing of a pregnant woman required compensation of 600 *solidi* (§ 19). The killing of an unborn child or a newborn baby before the bestowing of a name was to be compensated by damages in the order of 100 *solidi* (§ 20). Finally, in the last paragraph (21) there is mention of killing a freeman in his home. In this case compensation was 600 *solidi*.

The order for the enumeration of murder victims is fairly unclear. For it appears as follows: a freeman, Frankish courtier or free woman, three categories of Romans, unmarried girl, a mother, an old woman, a boy, a pregnant woman, an

---

396 Cf. R. Schmidt-Wiegand, *Fränkische und frankolateinische Bezeichnungen für soziale Schichten und Gruppen in der Lex Salica*, [in:] Nachrichten der Akademie der Wissenschaften in Göttingen, Philol.-Hist. Klasse nr 4 (1972), pp. 231–232; the word *trustis* comes from the Old Frankish *trôst* – 'backing unit'; cf. *Cap. leg.sal. add.* I, LXXI, where it is stated that compensation for the castration of an *antustrion* was 600 *solidi*, while for a free Frank 200, that is the equivalent of their *wergelds*.

unborn child or newborn baby and again a freeman. It seems, therefore, that the codifiers divided the victims of murder into three groups: men, women and minors. It is not, however, understood why between an old woman (§ 17) and a pregnant woman (§ 19) there is placed a long-haired (uncut) boy (§ 18). Similarly, for reasons unaccountable, in paragraph 5 next to the courtier there is included a free woman, which equally breaks up the division into the three mentioned groups. This paragraph suggests that compensation at a rate of 600 *solidi* was allocated for all free women. (Possibly, however, those encoding the law were not concerned in essence about all freewomen, but simply those from the families of court dignitaries.) Yet in the content of paragraphs 16, 17 and 19 it results that this rate was valid only in the case of women who could give birth as well as pregnant women. The common feature for all the mentioned categories of individuals is for certain the free legal status of the individuals. Freed persons and slaves are here absent, which is no surprise given that title XLI was entitled *On Killing Free People*.

We find some of the mentioned regulations in two other sections of *Pactus*: XXIV. *De homicidiis parulorum et mulierum* and LXVe. *De muliere grauida occisa* – although they are not always formulated in the self same way as in title XLI. In paragraph 1 section XXIV there is talk of the killing of a free boy 'under twelve up until the completion of the twelfth year'.[397] The blood money in this case was 600 *solidi*. This regulation recalls paragraph 18 of title XLI with the difference that there the mention is of a boy with long (uncut) hair (*puer crinitus*).[398] The *wergeld* rate was, however, the same. Paragraph 4 of title XXIV is therefore formulated almost the same as the said point 18 of title XLI.[399] Therefore it is seen that *puer ingenuus infra XII annos* and *puer crinitus* are two different categories even though their *wergelds* are identical.

In point 5 of title XXIV talks in turn of the beating to death of a pregnant freewoman. Compensation in this case was 700 *solidi*. This entry recalls to a degree paragraph 19 of title XLI, on the murder of a pregnant woman, in which, however, the damages to be paid were equal to 600 *solidi*. The difference is also in the formulation – in title XXIV, 5: *Si quis femina ingenua et grauida trabaterit* [...] *si moritur* [...]; in XLI, 19: *Si quis feminam grauem occiderit* [...]. Point 6. of title XXIV concerns the killing of a child while still in the mother's womb

---

397 *LSal.*, XXIV, 1. [...] *infra XII annos usque ad duodecim plenum* [...].
398 *Ibidem*, XLI, 18. *Si quis puerum crinitum occiderit* [...] *solidos DC culpabilis iudicetur*; K. Fischer Drew (*The Laws of the Salian...*, p. 45) considers that the difference in the rate of the *wergeld* between boys aged under and over twelve (600 and 200 *solidi*) resulted from the fact that the former, as defenceless, were to be protected from violence while the latter were acquainted with arms in the same way as grown men.
399 *Ibidem*, XXIV, 4. *Si quis puerum crinitum occiderit* [...] *solidos DC culpabilis iudicetur*.

or before a name was given, up to the ninth day after birth and was formulated like paragraph 20 of title XLI. In both cases the compensation was to be 100 *solidi*. The order of these two pairs of regulations in both titles suggest that they concerned the same categories of people. However, we are here dealing with an important difference in the value of blood money. The 700 *solidi* mentioned in section XXIV is, as it seems, the sum of two figures: the *wergeld* of a free woman able to have children (600 *solidi*) and the *wergeld* of a child (100 *solidi*). It is not clear why in title XLI the figure of 600 *solidi* is given.[400] The mentioned title LXVe contains another regulation on the killing of a woman with child, and also the child itself. Paragraph 1 states, in a similar way to title XXIV, that the *wergeld* of a pregnant woman is 600 *solidi*. However, it also contains another resolution, namely that if the foetus was male then the *wergeld* was to also be 600 *solidi*, just like in the case of killing a boy under twelve. This is a most significant difference in relation to the 100 *solidi* of titles XXIV and XLI. This note implied also that the lower rate was in force in the case of girls. In a further part of section XXIV, in paragraph 8 on the killing of a woman who had already given birth and in paragraph 9 talking of the self same crime against a woman following menopause, the *wergelds* were 600 and 200 *solidi* respectively. These regulations were repeated in points 16 and 17 of title XLI.

The structures of titles XXIV and XLI display a certain similarity. Paragraphs 4, 5 and 6 of title XXIV (on the killings of: a long-haired boy, a pregnant woman and an unborn child or newborn) were repeated in the same order in title XLI (§§ 18, 19 and 20), similarly points 8 and 9 of title XXIV (on a mother and a woman after menopause) were placed in title XLI as paragraphs 16 and 17. While the paragraphs from 1 to 15 (on a free Frank or a barbarian of Germanic origin, a courtier, three categories of free Romans and a free maiden) and paragraph 21 were formulated for the first time. This suggests that title XLI was to constitute a comprehensive regulation on the killings of various categories of free people. It does not, however, totally fulfil its function because in title LIV (*De grafione occiso*) there is a supplement to this list. It contains three points (1) on the killing of a graf (a *wergeld* of 600 *solidi*), (2) of a *sacebaron* or *obgrafio* (*puer regius*) – a *wergeld* of 300 *solidi* (3.) of a free *sacebaron* (600 *solidi* of blood money).[401] Additionally title XLVe contains regulations already earlier formulated in titles XXIV and XLI, and namely on the killing of a pregnant woman, an unborn child, girls to the age of 12, mothers under sixty years old

---

400 This difference has also been noted by K. Fischer Drew (*ibidem*).
401 A *sacebaron* was a royal court officer, *puer regius* and *obgrafio* were royal servants holding the office of *sacebaron* and count, hence their lower *wergeld* (K. Fischer Drew, *Introduction*, [in:] *Laws of the Salian...*, p. 46).

and women over sixty. Besides, in title XXXV (*De homicidiis seruorum uel expoliationibus*) there is talk of the killing of a slave man or slave woman commited by another slave. In this case the owner of the perpetrator and the owner of the victim 'shared' the murderer (or rather his market value) between themselves. There is a lack of a regulation on the killing of a *litus*, however, although in this title (paragraph 8) he appears as the murderer of a freeman.

We shall try to sum up and order this fairly tangled and not too coherent assembly of regulations, attempting to reproduce a picture of Frankish social structure that is as full and faithful to the regulations contained in *Pactus legis Salica* as possible. We shall order individual social categories according to the blood-money values, in a way differently from the method employed by the codifiers i.e., without a clear and coherent criterion.

To the group of people who were allocated the highest values of compensation for murder – i.e., 600 solidi, were:

> A free boy under twelve years of age or a long haired (uncut) boy (*crinitus*) – XXIV, 1 and 4, XLI, 18,
> A free mother or a woman breast feeding, up to the age of 60 – XXIV, 8; XLI, 16, LXVe, 3,
> A free pregnant woman – XLI, 19; LXVe, 1, according to XXIV, 5 – 700 *solidi*,
> A man *in truste dominica* – XLI, 5,
> *grafio* – LIV, 1,
> A free *sacebaron* – LIV, 3,
> A male foetus – LXVe, 1.

Wergeld at a rate of 300 *solidi* was given to:

> A *sacebaron* or *obgrafio* being half freemen of the king (*puer regius*) – LIV, 2.

The next group to have a blood money value of 200 *solidi* incorporated:

> A free Frank or Germanic man living according to Salian Law – XLI, 1,
> A free maiden or free girl under the age of twelve – XLI, 15; LXVe, 2,
> A free woman after menopause, upon reaching middle age or being over 60 years old – XXIV, 9, XLI, 17, LXVe, 4.

Compensation at a rate of 100 *solidi* was given to:

> An unborn child or newborn (female?) – XXIV, 6; XLI, 20.

In the case of the killing of slaves and slave women compensation was paid at a rate of 35 *solidi*.[402]

---

[402] *PLSal*, X, 3 and 6; the contents of title X is most unclear; point 3 talks of the stealing, murder or selling of another's slave, compensation was to be 35 *solidi*; K. A. Eckhardt gives two versions of point 6: according to the first the matter concerns the stealing or selling of a slave woman worth 15 or 25 *solidi* or a swineherd, a vineyard worker, a

From the above it results that in *Pactus* there were enumerated 16 different categories of person defined in relation to their sex, age, and social position. There is an absence, however, of 'ordinary' *liti*. Their blood money would have constituted a half of that of a freeman (i.e., 100 *solidi*).[403] The female sex was treated separately in the examined regulations. Its representatives were divided into 4 groups depending on the value of the sums that needed to be paid in the event of them being killed.

The criterion for division was the age of the victim, but first and foremost matters were decided by her ability for procreation. It also follows to emphasise that all the female categories of victim belonged to the group of free people. There is an absence here of slave women and freed individuals. Attention is also drawn by the fact that in the category of freemen there is no differentiation resulting from birth, but only with regard to the functions fulfilled (*antruscion, grafio, sacebaron*). This gives the impression that the Frankish codifiers were *not interested* in emphasising the differences in status amongst freemen, similarly to amongst various categories of person (free, freed, slave). This impression is heightened by the fact that the fines for bodily damage were allocated only for freemen. However, it is difficult to accept that other groups did not have their own compensatory rates.

In *Lex Salica* (texts D, E) the regulations on killing are to be found in six titles: XXXI, XXXII, XXXIII, LVII, LIX and LXIX. Title XXXI (*De homicidiis paruolorum uel mulierum*), regulated the problem of the killing of minors and women. Paragraph 1 concerned a free, long-haired boy and established his *wergeld* at a rate of 600 *solidi*. The beating of a pregnant woman resulting in death

---

smith, a miller, a carpenter, stable boy or other craftsman worth 25 *solidi*, while compensation was to be 72 *solidi*; according to the second version the matter concerned the stealing, murder or selling of a slave worth 25 *solidi* (11 professional categories are listed: *maior* (supervisor), *infertor* (estate manager), *scancio* (servant overseer), *mariscalcus* (equerry/groom), blacksmith, goldsmith, carpenter, vineyard worker, swineherd, household servant) – the compensation was to have been 35 *solidi*; the same in the case of slave women. K. Fischer Drew (*ibidem*, p. 47) writes about the amounts of compensation in the case of the stealing of slaves of various categories that are encoded in title X, 6–7, and which were from 15 to 25 *solidi*.

403 *PLSal*, XXVI, 1 *Si quis homo ingenuus alienum letum* [...] *extra concilium domini sui ante regem per denarium ingenuum dimiserit* [...] *solid C culpabilis iudicetrur.*; Cf. K. Modzelewski, *Barbarzyńska...*, pp. 73–74, 187–188; in title XLII, two Salian laws state that the *wergeld* of a free Frank killed in his home by a band (*contubernium*) equalled 600 *solidi*; paragraph 4 of the self same title determined that in the case of a *litus* the payment was to be a half of this amount (i.e., 300 *solidi*), analogically therefore if the 'ordinary' *wergeld* of a freeman was 200 *solidi*, then in the case of a *litus* it should have been 100 *solidi*.

(paragraph 2) brought about the need to settle a compensatory claim of 300 *solidi*. (In *Pactus*, XXIV, 5, in an analogical case, the *wergeld* was 700 *solidi*.). Paragraph 3 covered the death of an unborn child. For this act there was established compensation at a rate of 100 *solidi*. Title XXXII, being as if a continuation of the previous one, regulated the case of the killing of a mother, whose blood money was 600 *solidi*. Title XXXIII in turn talks of a woman after menopause, who had become middle-aged (paragraph 1) for whose killing the perpetrator was to pay 200 *solidi* and about a free maiden-virgin (paragraph 2), whose blood money was 100 *solidi*. (For comparison, in *Pactus*, XLI, 15, the *wergeld* for her was established at 200 *solidi*). Title LVII (*De homicidiis seruorum uel expoliacionibus*) contains only one paragraph of interest to us (1), but for all that extremely significant. For it concerns damages for the murder of a slave or slave woman commited by another slave, this being valued at 20 *solidi*. Finally title LXIX set the *wergelds* of a free Frank or Germanic man living in accordance with Salian Law (paragraph 1) – 200 *solidi*, the royal *antruscion* (paragraph 4) – 600 *solidi* and three categories of free Romans. As can be seen, texts D and E of *Lex Salica* in the three cases of pregnant women, free maiden-virgins and slaves differed significantly from the regulations of *Pactus legis Salicae* with regard to the rates of damages paid for killing. Attention is drawn by the fact that in *Lex Salica* D and E there occur 9 categories of victim, while in *Pactus* – 16 categories. Similar differences appear amongst the texts D, E and S. These latter, both in terms of the number of victim groups as well as the *wergeld* rates, more readily recall the regulations contained in *Pactus*.

* * *

The above analyses show that the picture of social structure was constructed in the Frankish codifications on the basis of a different principle than in the laws of the Frisians and Lombards. For there a key role was played by the hierarchical system of the various status categories. In the Frankish legal codes the differences in status were but one of the differentiation criteria. The sex of the victim of murder played a much more significant role than was the case for the Frisians and Lombards. It was taken into consideration even in the case of the death of unborn children. The division into a male and female population was one of the main elements constructing the picture of Frankish society. Another criterion was the age of the victims, which particularly in the female part of society (a girl under twelve or a free maiden, a mother, old woman) had a huge significance. However, this was equally important for the male population (male foetus, a long-haired boy or twelve-year-old, a free man). The differentiation with regard to social position was not of crucial importance. The significant majority of the categories of murder victim that were enumerated in Frankish codifications were ordinary freemen. These did not include *antruscions* or *grafiones*, free *sacebarons*, half-free *sacebaron* and slaves. The dominant grouping in this picture was that of ordinary freemen, divided according to the criteria of sex and age.

A comparison of the Frisian, Lombard and Frankish regulations show that we are dealing with three separate ways of shaping a model image of tribal social structure. These are their synthetic characteristics. The picture of the Frisian community in the light of title 1 (*De homicidiis*) was extremely homocentric, that is was reduced to the male part of the population as well as comprehensively covering all the status groups (the nobles, ordinary freemen, *liti* and slaves). This scheme omitted women, children and other groups. The ideological message derived from the title should be understood thus – the male part of the population as an entirety constituted, for those drawing up the law, the core as well as the root of the Frisian tribal community.

It follows here to note that this was an exception when viewed against the background of other Germanic codifications. No other code of tribal law omitted women in its model vision of society in such a radical way. This feature of *Lex Frisionum* is especially striking in comparison with the regulations of the Salian Franks.

The 'image' of the Frankish community, as we may observe in *Pactus legis Salicae* and *Lex Salica* was dominated by two groups – and namely women and children. From amongst the 16 categories of individual listed in *Pactus* women and children constitute 8 groups (4 for each). All of these categories simultaneously are numbered among the 'order' of ordinary freemen, which constituted the second dominant characteristic of the discussed picture. They created in total 9 status categories. Besides, a *wergeld* at its highest rate (600 solidi) was awarded to women and children in four cases out of seven.

They were allocated within the researched vision of society the same 'ranking' as dignitaries and officials. Against this background men – ordinary freemen – occupied a poor position in relation to the value of *wergeld*, as equally in terms of the number of subcategories within the male population. Attention is drawn by the omission of the freed, as well as women from the groups of dignitaries and officials. In terms of the place occupied by women and children within the picture of social structure, Frankish codifications represent an exceptional solution.

The rules of *Edictum Rothari* are equally a specific case. This results first and foremost from the fact that this is the only Germanic codification (not counting the *Liutprand Laws*), in which the picture of social structure was contained in a register of damages and not in regulations establishing *wergeld* rates. This picture combines within itself certain features of the two presented above. For, on the one hand, it has, given its systematic nature, a comprehensive character – for it incorporates freemen, *liti* and slaves. While on the other hand, there are considered within it, although not fully, women (*aldiae* and slave women).

* * *

In summing up, we can state that although the social structures of particular Germanic peoples were presumably similar with regard to the basic characteristics, the legal texts, however, 'modelled' their image in a varied way, in accordance with the different ways particular tribal codifiers visualised their own people. Here are several examples which show what this phenomenon involved. Let us start from the group of freewomen. In each society there were girls under the age of twelve, maidens, mothers and women after menopause, but only in Salian and Ripuarian laws were these incorporated into 4 separate categories. In the legal codes of other peoples there appeared maidens and married women (Bavarian, Saxon, Visigoth and others) but there was an omission of young girls and old women as separate legal groups. Another example are slaves. Presumably everywhere they constituted a differentiated socio-legal category but, for example, in Lombard law (*Edictum Rothari*) there are listed 9 subcategories of slave, in Burgundian law 3, while for the Frisians, Saxons, and Bavarians slaves were classified as merely a single group. Finally freemen: presumably in all the researched tribes this was a varied group (in relation to wealth, birth or honour), however in the particular *leges* they were classified in a different way. According to the Visigoths they were a homogeneous group (a single *wergeld* rate), in turn in the law of the Alamanns there were distinguished three categories of freeman, in a similar way to the Burgundians. However, according to Lombard law even though the free were a single category there were different *wergeld* rates (depending on *qualitas personae*).

As we have already recalled the driving idea shaping the classification of the social groups contained in the legal codifications of the individual Germanic peoples was the concept of a modelled social order, though it was not the intention to depict within this all groupings or categories of person. Those codifying the law consciously chose only those which in their opinion played a significant role in their vision of community order or those whose situation appeared at the time of conducting the codification to be in particular need of regulation (resulting from actual needs). The concept of social structure was therefore always a construction created by the codifiers in accordance with their convictions. Knowledge about the social structure of their people was used by them in their need to formulate rules regulating killings and bodily violations. Therefore they utilised this selectively, hence the constructed picture of society as a rule is in each of the researched texts more or less fragmentary.

Textual Presentations of the Human Body and Model Depictions of Social Structure

Both the mechanisms used for composing the groups of provisions pertaining to crimes against the body and the modes of perceiving the body which are bound up with them were discussed in detail in Chapter II, so we will confine ourselves here to recalling some of the conclusions that follow from what we have established. The most important issue we will deal with in this part of the present chapter is the relationships between textual presentations of the human body and model depictions of the social structure which we can observe in the codes of the laws of separate Germanic peoples. What is of particular interest to us is how that presentation of the human body was incorporated into the classification of social groups adopted in a given codification. For the investigated texts sometimes differed from each other quite markedly in this respect.

The relationship between social structure and an individual's body is outlined in the *Edictum Rothari* in a manner that is particularly interesting from our topic's perspective. The issue of the depiction of the Lombard community has already been discussed above, so let us move on to the issue of the depiction of the human body. As we remember, this was included in the regulations pertaining to crimes consisting of breaches of bodily inviolability and integrity, especially that of men. Let us now investigate the manner in which the human body model was used in relation to the separate status groups that helped to shape the depiction of Lombard social structure contained in the *Edictum Rothari*. Free people – more strictly speaking freemen – were the primary social category with regard to which the Lombard codifiers formulated the provisions regulating crimes against the body. These provisions are contained in titles 41 to 73, as well as in 377 and 382 to 384. A total of 55 crime instances are listed in these, which relate to 26 different parts of the body.[404] Let us remind that these were listed in order of head to toe (see: titles 45 to 73). Each of the breaches of bodily inviolability or injuries to the body of a free man mentioned in these provisions caused a need for the perpetrator to pay out a particular compensation sum.

While keeping these observations in mind, let us take a similar approach to the investigation of the regulations pertaining to the other two groups making up the depiction of the structure of the Lombard community, i.e., the *aldii* and the slaves. As was mentioned earlier, the investigated provisions pertaining to *aldii*

---

[404] Titles 41 to 43 pertained to crimes against the whole body or a particular location on the body; in the case of 44, the face, then from 46 to 73, injuries (respectively) to: the skin on the head, skull, face, an eye, nose, cheek, front teeth, side teeth, an arm, the torso, a thigh, shin, hand, thumb, index finger, middle finger, ring finger, little finger, foot, big toe, 2nd, 3rd, 4th and 5th toe, while title 383 relates to the beard and hair.

were included in titles 77 to 102, as well as in 377 and 383. These provisions contain a total of 43 instances of crimes against 26 body parts.[405] The same applied to the crimes against the body of a slave in titles 103 to 126, and also 377 and 383.[406] The regulations pertaining to *aldii* and slaves contain, for each of these categories, separate compensation sums for particular criminal acts. This was a factor differentiating the position or status of both of these groups from each other, and also in relation to free people. However, it should be strongly stressed that in the case of all three social categories, an almost identical order for listing crimes was adopted, based on a conventional image of the human body presented from the head to the feet.

Such a mode of referring to the set of human body members is very important for understanding its role as an indicator of status for different social groups. It also has certain significant consequences. Let us try to explain the significance of the described lawmaking practices. The first observation that arises in relation to the problem being discussed here of the relationship between the social structure and the human body is as follows: in the *Edictum Rothari* (and also in other Germanic codifications), the set of members of a free man's body formed a kind of pattern, which was reproduced in relation to other status categories (free women, freedmen or slaves). A free (common) man, as a representative of his category, represented, in the light of the discussed provisions, a kind of centre or axis of the tribal community. His body served, at the same time, as a model of the human body in general. Secondly – and this is a conclusion of fundamental importance and with wide-ranging consequences – the presented practice of the Lombard codifiers indicates that they treated the body as a factor which assimilated or simply set representatives of the aforementioned three social groups (free men, *aldii* and slaves) on an equal level. Irrespective of the differences in the levels of compensation which appeared between them, the freeman, *aldius* and slave were essentially the same in a physical sense.[407] The body was a common generic characteristic. Assuming the Lombard codifiers conceived of human body members as a manifestation of a man's humanity, it

---

405 There is mention there of damage to the skin on the body and head, on the skull, face, an eye, nose, ear, cheek, front tooth, side tooth, arm, hand, thumb, index finger, middle finger, ring finger, little finger, knee, foot, big toe, toe numbers 2, 3, 4 and 5, torso, beard and hair.

406 These applied to injuries to: the skin on the head, skull, face, an eye, nose, ear, cheek, front tooth, side tooth, arm, thigh, the trunk, a shin, hand, thumb, index finger, middle finger, ring finger, little finger, big toe, toe numbers 2, 3, 4 and 5 and the skin.

407 E.g., compensation for a facial injury amounted, in the case of a free man, to 16 *solidi* (title 54), 2 *solidi* in the case of an *ald* and educated slave (title 80), and 1 *solidus* (title 104) in the case of a rural slave (title 104).

can be supposed that they treated those belonging to *aldii* and slaves in exactly the same way. This observation would seem to lead to the conclusion that the body (or body members) had a 'humanising' effect, i.e., they served as an indicator for membership of the world of people, and this applied equally to all three of the aforementioned social groups.

The above viewpoint, although it may appear banal and obvious, in actual fact is not in the least that. For there are certain arguments attesting to the fact that 'barbarian' lawmakers failed to perceive slaves as people. Karol Modzelewski, when reflecting recently on the issue of the legal situation of slaves in the codifications that are our focus of interest, presented the thesis that, being deprived of legal subjectivity, they were the objects of someone else's property law.[408] The author refers to rules in the *Lex Salica* (title X, 1) and *Pactus legis Alamannorum* (XXIX) in which mention is made of the death of a slave (caused by an animal) and the stealing of a slave, in effect equating them with animals.[409] K. Modzelewski comments on the latter instance as follows: 'Here the slave was contrasted, using the unreflected upon obviousness of colloquial language, with a human being and placed in the same category as "some animal"'.[410] The author goes on to establish that: 'a slave was not regarded as being a subject of the law in any situation: either when he fell victim to a crime or in a situation when he actually committed a crime'.[411] Slaves also could not be witnesses at court.[412] The Germanic laws also failed to recognise slaves' family ties.[413] These are arguments with which it would be difficult not to agree. Yet the question arises of how to reconcile the viewpoint of the cited author on the exclusion of slaves from human society with the earlier quoted entries from the *Edictum Rothari*.

---

408 K. Modzelewski, *Barbarzyńska...*, p. 173.
409 *LSal*, X, 1. *Si quis seruum aut ancillam alienam, caballum uel iumnentum furauerit* [...] *solidos XXXV culpabilis iudicetur*; *PLA.*, XXIX, 1. *Si caballus, boves aut porcus hominem occiderit, totum viregildum solvatur*; 2. *Si servus fuerit aut quemvis pecus, medium precium solvatur.*
410 K. Modzelewski, *Barbarzyńska...*, p. 174.
411 *Ibidem*, p. 176. the author here refers to the provisions of the law of the Turingians (Title LIX) and the Swedish *Westgötalag* (*Af mandrapi*, 1, 4), as well as the *Edictum Rothari* (Titles 370, 371 and 373).
412 *Ibidem*, p. 179, cf. *PLSal*, XXXIX, 2; as well as *Prostrannaja Prawda*, [in:] *Prawda Russkaja*, vol. I, ed. B. D. Grekow, Moskwa–Leningrad 1940, Title 85: *Budiet li posłuch chołop, to chołopu na prawdu nie wyłaziti.*
413 *Ibidem*, s. 182; por. *LBaiw*, VIII, 10. *Si quis cum manumissa, quod frilaza vocant, et maritum habet, concubuerit, cum 40 solidis componat parentibus vel domino vel marito eius*; 12. *Si quis cum ancilla marita concubuerit, cum 20 solidis conponat domino.*

This is not the only doubt regarding the discussed position to emerge from a reading of the oldest Lombard law book. What is of concern here are the contents of title 383 of this code. There is mention in this of a crime involving tugging someone by his beard or hair during a fight. The following compensation sums were prescribed for perpetrating this act: 6 *solidi* for a freeman and the equivalent of the fee for a single blow (*sicut pro ferita una*) for an *ald* or educated or rural slave, i.e., 1 *solidus* and ½ *solidi* respectively.[414] We are dealing here with a particular kind of crime against the body. Most crimes mentioned in the analysed compensation tariffs involve violation, to a greater or lesser extent, of the body substance. Yet title 383 applied to violations of bodily integrity which presumably failed to cause physical harm. The crime therefore entailed perpetrating an act of violence on the injured party that encroached on his dignity.

The following question arises at this point: How can we explain the fact that the compensation system was binding in this case for all social groups, including slaves? According to K. Modzelewski, a slave, in the light of the law, was not a human being. So how was it possible to breach his bodily inviolability and violate his dignity? Even if it is taken into account that the compensation sum stipulated in the case of a slave was paid out to his master in recognition that his property rights had been violated, this fails to explain why the cited entry came into being. For it only made any sense if it was assumed that the slave possessed personal dignity. Otherwise, it would not have been possible to violate this. It is also difficult to accept the explanation that as a result of the described crime, it was the owner of the slave, rather than the slave himself, who suffered damage to his honour. This would be an exceptionally circuitous thought construction. For it would then be necessary to accept that the dignity (or other personal possessions) of the master were also violated when, for example, the tail of his horse or cow was tugged, which is unconfirmable from any of the barbarian law codes that are known to us. This is precisely what K. Modzelewski is suggesting when he talks of placing slaves and animals in one legal category. While it can be understood that the master would suffer a (material) loss as a result of the violation of a slave's body substance, it is difficult, in the case of beard or hair tugging, to speak of the destruction of someone's property. It is not possible to offend someone who has no dignity. Yet dignity, in the light of the barbarian laws, was only possessed by human beings. It therefore appears that in the Lombard code, slaves must have been regarded, at least as far as bodily inviolability was concerned, as people.

There is one more argument in favour of the thesis that the body, in the conviction of the creators of the barbarian laws, was a feature shared by people be-

---

414   Cf. *ERot*, 77 and 125.

longing to different socio-legal groups, including slaves. There are regulations pertaining to corpse and grave defilement. The laws of the Alamanns mention the digging up of the corpses of a free man and free woman, and also those of a male and female slave.[415] The Salian law (*Pactus* and *Lex*) even mentions the stealing of the corpses of a free man and someone else's slave.[416] In the light of the cited provisions, the acts presented in them were treated as crimes both in cases where the objects of crime were free people and when they were slaves. The law protected the human corpse no matter what legal status a person possessed during his lifetime. The fact that a corpse (or dead body) was in a legal sense a feature common to both free man and slaves was explicitly expressed in the *Lex Baiwariorum*, in a provision relating to the burying of the dead: 'If anyone anywhere finds a dead person (a corpse) and buries it out of humanity, so that it is not contaminated by swine or torn apart by wild animals or dogs, no matter whether a freeman or slave, and it is subsequently discovered, the person who buried it, if he so wishes, may demand from the family, or from the slave's master if it was a slave, 1 *solidus* in payment'.[417] Particularly noticeable in this sentence is the statement that the deceased was supposed to be buried *humanitatis causa* ('out of humanity') or to use contemporary notions – 'out of humanitarian impulses'. This can be understood in two ways – the burial was performed

---

415 *LAl*, XLIX (B: *De eo, qui liberum de terra effodierit*), 1. *Si quis liberum de terra exfodierit, quidquid ibi tullit, novigeldos restituat, et cum 40 solidis conponat. Femina autem cum 80 solidis, si de terra exfodierit; res autem, quod tullit, furtiva componat.* 2. *Si servum exfodierit de terra cum 12 solidis conponat; et ancilla similiter*; cf. also, *PLAl*, XVI; 3. *Et quicumque mortuo, tam occiso quam qui sua morte morit, aliquid tollatur aut involatur, de fossa, ubi reponatur, exfoditur et expoliatus fuit, quod ibi tullit, reddat et 80 sol. solvat.*

416 *PLSal*, XXXV. (*De homicidiis seruorum uel expoliationibus*), 6. *Si quis seruum alienum mortuum in furtum expoliauerit et ei super XL denarios ualentes tulerit, mallobergo theofriomosido, MCCCC denarios qui faciunt solidos XXXV culpabilis iudicetur;* LV. (*De corporibus expoliatis*), 1. *Si quis corpus occisi hominis, antequam in terra mittatur, in furtum expoliauerit cui adprobatum fuerit, mallobergo chreomosido hoc est, MMC denarios qui faciunt solidos LXII semis culpabilis iudicetur*; *LSal*, (S) XX. (*De eo qui mortuum hominem expoliauerit*), *Si quis hominem mortuum, antequam in terra mittatur, furtu expoliauerit IVM denariis qui faciunt solidos C culpabilis iudicetur.* LXVI. (*De homicidiis seruorum uel ancillarum*) 6. *Si quis seruum alienum mortuum per furtum expoliauerit et spolia ipsa plus quam XL denarii ualeant, MCCCC denariis qui faciunt solidos XXXV culpabilis iudicetur.*

417 *LBai*, XIX. (*De mortuis et eorum causarum*): 7. *Et si aliquis a quolibet mortuus fuerit repertus, et eum humanitatis causa humauerit, ut neque a porcis inquinetur nec a bestiis seu canibus laceratur, liber sit an servus, et postea repertum fuerit, et ille eum humauerit, si requirere voluerit parentes vero illis, solvant solidum ei unum aut dominus servis, si servus fuerit.*

on account of the humanity (or human feelings) of the person burying the corpse or due to the humanity (or human dignity) of the person being buried. These interpretations, rather than being mutually exclusive, tend to corroborate each other. This is clear from the next statement, in which the corpse was to some extent contrasted with animals, for the burial was supposed to protect the deceased from them. Leaving a human corpse, *sit liber an servus*, to the mercy of animals would be inhuman. This case shows that free people and slaves were set on an equal plane by the law.

The practice entailing treating the human body as a feature common to men belonging to the three main status categories (free men, freedmen and slaves) is not a feature that is unique to the *Edictum Rothari*. We also see it in the *Lex Baiwariorum*. The provisions of interest to us in the Bavarian codification that speak of crimes against the body were included in three separate chapters: IV. *De liberis, quomodo conponantur*, V. *De liberis qui per manum dimissi sunt liberi, quod frilatz vocant* and VI. *De servis, quomodo conponantur*.[418] The content and layout of these chapters also outlined a vision of the social structure of the Bavarian people, which is adopted by those recording this law code. Chapter IV, which is devoted to free men, contained descriptions of 50 crimes against 28 body parts.[419] Chapter V, pertaining to freedmen, appears quite exceptional, for it only contains 18 crimes against 17 body parts.[420] Finally, Chapter 6, which speaks of slaves, included descriptions of 30 crimes against 27 body parts.[421] The structure of the provisions being discussed here is similar to that which we can observe in the *Edictum Rothari*. In the codification of the Bavarians, the principle held that an image of a free man's body served as a model which was reproduced in the case of freedmen and slaves. The same order was also preserved for listing body parts and crimes entailing their violation. Another similarity between both compensation tariff systems consisted in the use of graduat-

---

418 *LBaiw*, IV, 1–16 (pp. 316–328); V, 1–8 (pp. 338–341), VI, 1–11 (pp. 342–346).
419 This chapter includes 30 items, in which the following violated objects are mentioned (in order of appearance): the skin on the body, any vein/artery, skin on the head, bone of the skull, bone of the arm, meninx, the innards (abdominal cavity?), an eye, hand, foot, thumb, index finger, little finger, middle finger, ring finger, arm above and below the elbow, nose, ear, lower eyelid, lower lip, upper eyelid, upper lip, molar tooth, genitals, another tooth, leg at the knee.
420 Chapter V contains only 9 items, in which the following violated objects are discussed: the skin covering the whole body, any vein/artery, skin on the head, bone of the skull, bone of the arm, meninx, the innards, an eye, hand, foot, thumb, index finger, little finger, middle finger, ring finger and legs at the knee.
421 Chapter VI is comprised of 12 items, which contain all the objects mentioned in chapter IV, with the exception of the genitals, and in the same order of appearance.

ed compensation rates for the same crimes against each of the status categories.[422]

The titles discussed in the *Lex Baiwariorum* also display certain differences in relation to the oldest law of the Lombards. The first of these – an order for listing body members that is not modelled on the head to foot order – was dealt with in Chapter II. The second difference, more important from our perspective, manifests itself in the fact that title V, which is devoted to freedmen (*frilatz*) contains significantly fewer body parts than that which pertains to freemen, but fewer also, surprisingly, than the one regulating the position of slaves. This peculiarity in the codification of the Bavarians is difficult to explain, especially if it is viewed in the context of the resolution adopted in the *Edictum Rothari*, whereby differences between the three 'states' are minimal in this respect. Also surprising is the fact that this reduced list of body parts (and also of crimes) appeared in the case of freedmen rather than slaves. This would have agreed with the principle of hierarchical status prevailing in the Germanic *leges* (not only the Lombard laws) that was expressed through the differentiation of compensation levels.

The impression may be gained that, paradoxically, slaves were treated better in this respect in the *Lex Baiwarorum* than freedmen, who were after all granted damages that were at twice the level of those granted to slaves. Leaving aside the question of how to explain this unclear issue, let us underline that, despite the indicated differences in the items that are of particular interest to us, the codification of the Bavarians displays marked similarities to the *Edictum Rothari* in terms of structure and content. The clear convergence between the images of the human (more precisely: male) body in the chapters pertaining to free men and slaves convinces us that the Bavarian legislators assumed a kind of unity prevailed between all social categories as a result of the physical features common to them all.

The clauses in the code being discussed are reminiscent of the oldest Lombard edict in one other respect. There is a provision here speaking of a crime known as *infanc* in Old Bavarian, which was rendered in Latin as *manus inicere in aliquem*[423], i.e., raising a hand to or attacking someone. This act, although it undoubtedly had the makings of a physical assault, was nevertheless character-

---

422 E .g., in the law of the Bavarians, the composition for gouging out an eye amounted to: 40 *solidi* in the case of a free man (IV,9), 10 *solidi* in the case of a freedman (V, 6) and 5 *solidi* in the case of a slave (VI, 6). The same principle applied to all other crimes of this type.

423 Cf. *LBaiw., Glossarium verborum vernaculorum*, p. 487; IV, 3: *Si in eum contra legem manus inicerit, quod infanc dicunt* [...],V, 3: *Si in eum contra legemmanus inicerit, quod infanc dicunt* [...], VI, 3: *Si in eo contra legem infanc fecerit* [...].

ised differently to the other crimes. For the outcome of this act, i.e., physical harm, was not specified. It appears that this act involved the use of violence or the threat of its use. It is not therefore a violation of the body substance that is at issue here, but rather an attempt to breach the inviolability of the injured party. But the point here is that the regulations relating to this crime can be found in all three of the chapters being discussed, which means that – they applied to free men, freedmen and slaves. *Infanc*, as a crime that employed violence but failed to cause physical harm yet quite possibly entailed a breach of bodily inviolability, has a similar purport in the *Lex Baiwariorum* to the beard or hair tugging in the *Edictum Rothari*. For it indicates that those recording the laws of the Bavarians were anxious to protect the non-material attributes bound up with the human body, and this protection took men of all status categories into consideration. This means that they must have recognised that slaves also possessed these attributes, but to a lesser extent than free people and freedmen, all of which is evident from the differences in compensation amounts.

A similar way of thinking about the socio-legal dimension of human corporeality can also be observed, although on a somewhat smaller scale, in the codification of the Ripuarian Franks – the *Lex Ribuaria*. When viewed through the prism of the compensation system for crimes against the body, the structure of the male portion of this population incorporated two groups: freemen and slaves. Apart from this, on the basis of the *wergelds*, it is possible to identify the king's people (presumably those who had been freed) – both men and women (*homines regis, feminae regis*)[424], ecclesiastical people (*homines ecclesiastici, feminae ecclesiasticae*)[425], people *in truste regia*[426], free married women (*feminae Ribuariae*), free maidens (*puellae Ribuariae*) and women over forty.[427] In the case of free men, 20 different crimes against the body and 12 injured body parts are listed.[428] With regard to slaves, the Ripuarian codifiers indicated 12 acts and 8 objects of crime.[429] Although the difference between freemen and slaves is clear, it is not, however, radical in nature. Moreover, as is the case in the Lombard and Bavarian codifications, the image of the structure of a freeman's body served as

---

424   *LRib*, IX i XIV, 1.
425   *Ibidem*, X i XIV, 1.
426   *Ibidem*, XI.
427   *Ibidem*, XII, XIII, XIV, 2.
428   These provisions were contained in titles I to VI. They cover the following body parts: the skin covering the whole body, any bone, the torso, an ear, nose, eye, hand, thumb, index finger, foot and the genitals.
429   Titles XIX to XXVII cover: the skin covering the whole body, any bone, an ear, nose, eye, hand, food and the genitals.

a model, which was reproduced in a slightly truncated form to apply to slaves, with the same order of body members being maintained.

The cited examples of resolutions adopted in the *Edictum Rothari*, *Lex Baiwariorum* and *Lex Ribuaria*, convincingly demonstrate, though to a varied extent, that we are dealing in these codifications with a clear link between the textual presentations of social structure and of the human body. In the aforementioned codifications, a model of a freeman's body was reproduced for other status groups, including slaves. The *Edictum Rothari* also incorporated some women, notably freedwomen and slaves. In the *Lex Baiwariorum*, textual presentations of a freeman's body served as a template for the paragraph devoted to free women.[430] The other two categories of women were not regulated in this way.

* * *

Analytical review of the model presentations of the tribal communities and research on the relationships between these notions and the perception of the human body incline investigation into the relations between these elements. What is of particular concern here is the way in which sets of body parts were used to create the aforementioned provisions relating to different social groups. On the basis of previous observations, it is fair to say that we are dealing in individual codifications with several different resolutions in this area.

The law books of the Salian Franks and Lombards (especially the *Edictum Rothari*) represent examples of different ways of describing the relationship between model presentations of the body and society. In the *Pactus legis Salicae*, the depiction of tribal structure was contained in the provisions stipulating compensation amounts for murdering representatives of different social groups. A portrayal of the human body, or rather, as our previous findings would indicate, a representative list of its parts, was only compiled in the case of freemen. Moreover, the codifications of the Salian Franks included provisions defining responsibility for breaches of the inviolability of some parts of a freewoman's body (fingers, hand, arm, breasts, hair). Such a manner of formulating regulations pertaining to crimes against life and health may be regarded as a specific type of relationship between textual presentations of social structures and the body.

A completely different resolution of this issue can be observed in the laws of the Lombards. In the *Edictum Rothari*, a model presentation of the construction of society was included in the provisions pertaining to violations to the bodies of representatives of the three social groups. These regulations incorporate: freemen, freedmen and freedwomen and male and female slaves, i.e., almost all the

---

430   *LBaiw*, IV, 30.

groups in the model depiction of Lombard society. Alongside this existed provisions which stipulated the *wergelds* for *aldii* and eight groups of slaves. There was a separate provision in the *Edictum* which regulated the cases when a freewoman was murdered. The depiction of the human body in the Lombard codification played an incomparably larger role than that in the Salian codification. It seems, therefore, justified to state that in *Edictum Rothari* or more broadly within Lombard law there was adopted a specific (and one completely different than for the Salians) type of relationship between the two structural depictions herein discussed.

The difference between both models (the Salian and Lombard) is dual in nature. Firstly, in the codifications of the Salian Franks, textual presentations of the human body did not play any role in conveying an outline of the social structure, while in the Lombard codification they were of key importance in this respect. Secondly – one connected to the first – in Salian law the human body as a comprehensive structure was reduced to a freeman's body, while in Lombard law, this image was expanded to incorporate the other social groups (with the exception of freewomen).

The differences between the Salian and Lombard models demonstrate, in an extremely explicit manner, differences in the way the relationship between the presentation of the human body and the depiction of social structure within a given society were shaped. But there are other variants of this relationship. Of these, the model contained in the *Lex Frisionum* deserves most attention. As is the case with Salian law, the depiction of social structure is included in the title pertaining to compensation for murder. But this only incorporates men. The depiction of the structure of the human body in the title devoted to crimes against the body uses the free man as a prototype, but this was stretched to incorporate two other groups – noblemen and freedmen. It omits slaves, but also every category of women. In the Frisian codification, the title containing a list of body members represented, as regards the presentation of social structure, an almost faithful repetition of the concept contained in the title speaking of killings. The role of both regulations was therefore equivalent. The relationship between the depiction of the construction of tribal society and that of the human body is clearly outlined in this case.

Other codifications of the Germanic peoples display features which are akin to the three models described here. In the law of the Bavarians, a model for social group structure is evident both in the provisions devoted to crimes against the body and in those pertaining to murders. Both sets of regulations mainly applied to the male population. But they also contain provisions speaking of double compensation payments and *wergelds* stipulated for freewomen. Generally, however, the depiction of model tribal structure in the *Lex Baiwariorum* was very clearly linked to the notion of the human body. In this sense, the Bavarian resolutions were closest to the 'Lombard' type.

The Human Body and the Hierarchical Structure of Social Groups

The provisions investigated in this chapter, which delineate the scope of responsibility for killings and crimes against the body not only contained what we have called a classification of social groups within a tribal community, but also expressed the convictions of individual codifiers about the social order. This phenomenon is clearly evident in the provisions pertaining to the male portion of tribal societies, although, as we saw, it is also reflected in regulations speaking of women. The social organisation of the investigated peoples relied, in the light of the above presented normative entries, on a hierarchical structure of different status categories. In one Germanic codification after another, we are dealing with different variants of this hierarchical structure. In the law of the Franks, it incorporated dignitaries, free commoners and slaves; in the laws of the Lombards and Bavarians, free people, alds and slaves; in the codifications of the Alamanns, first, middle and subordinate freemen, as well as *liti* and slaves; and in the law of the Frisians, noblemen, free commoners, lits and slaves.

The principle of inequality of legal status for different social groups, which was expressed, for example, through differentiation of compensation levels for murder and crimes against the body, became a foundation of the social order. The notion of a legal status hierarchy defined the depiction of the structure of every tribal community included in the investigated codifications. The compensation sum for murder or bodily violation expressed any given individual's place in society. The regulations describing the recipient of that payment, i.e., the 'beneficiary', clearly outlined the relationships between various status groups. This can be seen in the provisions speaking of killings or violations to the bodies of slaves and freedmen, which often stipulated that the compensation in these cases was completely (in the case of slaves) or partially (in the case of freedmen) payable to their masters.

The most obvious and indeed important manifestation of the relationship between the human body and the social order that emerges from the provisions of the barbarian *leges* being investigated here is that everything, even a minor 'physical object' (e.g., a single tooth, a finger part, a toe) was treated in them as a component of that social order. This was the case because every body part was assessed from the point of view of an individual's legal and social status, while status differences denoted differences in compensation levels. Legal status was in some sense recorded on the human body. An individual's set of body parts became an element in the concept of social order. Separate body members or fragments were not only legal objects but also their set as a whole was incorporated into the system of power relations (based on superiority and inferiority and hierarchies) which, in the light of the investigated codes, existed between social groups.

The attitude of the creators of the written barbarian laws to the human body was characterised by a certain paradox – on the one hand, the body was treated as the specific attribute of an individual yet on the other, it was presented as an object devoid of individual characteristics. The Germanic laws did not include regulations describing the principles dictating the establishment of compensation levels, thanks to which it would have been possible to adjust them to the life circumstances of a particular victim. In the light of the investigated provisions, compensation sums were not meant to be determined on the basis of negotiations between the injured party and the perpetrator, with every crime being treated separately. Instead they were determined by codifiers on behalf of separate social groups. These rates were 'averaged out' or standardised. There was no possibility of changing them by raising or lowering them according to circumstances. Quite clearly, the intention of those recording the laws was neither to ensure that levels of compensation should be the outcome of an investigation into the effects of a bodily violation on the existence of a particular injured party nor that the specific circumstances in which the crime occurred should be taken into account.

A compensation estimate for the violation of a given body part was derived from the socio-legal status of its owner and not its personal quality. A compensation sum therefore expressed, first and foremost, a person's position in society. The compensatory function, i.e., assessing compensation for a loss or other injury caused by a bodily violation or injury, would in fact seem to be secondary to the function of assigning status. If the opposite were the case, the standardisation of compensation payments would surely not have been possible. The legislators, by assigning permanent and uniform compensation rates to a given social group for particular injuries, assumed that there were no physical differences between the people within its limits. The bodies of nobles, free commoners or freedmen, as regards levels of compensation, were therefore reduced to uniform models.

As far as compensation tariffs were concerned, the human body possessed the value appropriate to the representative of a particular 'status' group. It was, therefore, assessed not only as the attribute of an individual but also as a feature of the group to which it belonged. When we read – in the chapters of the Germanic laws devoted to bodily violations – of a severed hand, struck out eye or broken bone and about the compensation payments for these crimes, we are not in fact dealing with a textual reflection of the actual members of a specific person, but with a 'status' hand, eye or bone. In this way, the body of an individual person or, more precisely, even minute fragments of this body, were harnessed into the idea of social organisation expressed by the law. Anything that was individual or private was presented in the language of legal norms as a component of the tribal public order. It seems that this is a mode of understanding (and also of acting) typical of societies in which the boundary between what is private or personal and what is public was fluid or quite difficult to grasp. The Germanic peoples of the early Middle Ages certainly displayed this characteristic of social life organisation.[431] In the light of the barbarian *leges*, the human body was at the same time an individual and public object. Both these aspects were so tightly interlinked that it is difficult to investigate them separately. Any attempt to separate them would appear to be futile or even inappropriate, because it would contradict the assumptions on which the analysed provisions are based or more broadly – certain convictions relating to society, the individual and the human body.

Man's Body and Affiliation to the Society

The relationship between a man's body and social structure may be still viewed within the barbarian *leges* from a different perspective. We have ascertained above that in some codes the body (particularly the male) in being the object of crime was perceived as a common feature (specific) of representatives of various status groups (freemen, *litus* (freedmen) and slaves). The human body so understood was in its own way a determinant of affiliation to human society – it designated its boundaries. Beyond this boundary was the world of animals. Animals, something taken into consideration by the creators of the legal texts, which equally possessed bodies. It is worth looking, within the context of our ascertainments with regard to the human body, at the way the creators of the said codes perceived the bodies of animals. A comparison of the regulations on animals and people will broaden, I think, our knowledge as to the body of man as a determinant of human social affiliation in the codes of the individual barbarian peoples. Of particular importance here is the question of the way an animal's body was treated in law on the one hand, while on the other that of slaves. According to our hypothesis slaves constituted in this regard a 'border' category of society – for they were situated as if on the edge of the human and animal world. The viewing of the body as the subject of legal regulations has, therefore, a significant meaning in determining the legal position of slaves, and particularly their social affiliation.

---

431 Cf. M. Rouche, *op. cit.*, pp. 420 ff.

In talking about the 'world of animals' attention needs to be paid to the differentiation conducted by Germanic codifiers. In *Edictum Rothari* there appears a division into wild animals (*ferae*), totally belonging to the natural world and being beyond the control of man, and domesticated animals (*animales*), incorporated into the world that lies under man's control.[432] The regulations in *Edictum* bring us closer to what its authors understood by the concept *animales* – and namely horses, cattle, dogs and swine. For these were the most typical domesticated animals. Their number may, obviously, be extended at least by poultry and sheep, as equally by the inclusion of other somewhat less commonly met species of domesticated animal.[433]

The division into wild and domesticated animals has significant consequences from the point of view of the research matter, consequences which we shall deal with in greater detail. Before we undertake this we shall state initially in what way the codifiers viewed domesticated animals as a subject of legal regulations. In title 326 of *Edictum* there is talk about a situation whereby a horse, ox, pig or dog inflicts some kind of bodily harm or kills a person (*homo*). The fines or *wergeld* for the damage inflicted was to be paid by the animal's (*animalis*) owner. Following the settlement of compensation the hostility (*faida*) between the victim and the owner was to cease – 'for [it] had been caused [by] a dumb thing (*muta res*), not (brought about it) the action of man (*hominis*)'.[434] The said 'dumb thing' was obviously one of the above mentioned animals, which was contrasted with man.

It is presumably no coincidence that the victim as a result of the said event was classified by a general term, one untypical for Germanic *leges*, through the use of the word *homo*. For one may surmise on this basis that the authors meant in this case man in general and not a representative of a particular status group. On the one hand we would therefore have domesticated animals defined as

---

432 Cf. *ERot*, 309 (*De fera*), 311–313, and particularly 314: *Si cervus aut qualebit fera ab alio homine sagittata fuerit* […]; in the translation of *Edictum* into Italian the word *fera* was rendered by *animale selvatico* 'forest animal'; see on the subject of the relations of the Franks and Burgundians to wild and domesticated animals within the context of hunting M. Rouche, *op. cit.*, pp. 487 ff.

433 *Ibidem*, 327. *Si quis praestitum aut conductum caballum aut bovem aut canem aut quolibet animalem habuerit*; Cf. also, *ibidem*, 315. *De cervo domestico* (On domesticated deer); 317. *De aves domesticas* (On domesticated birds); 322. *De canis incitatus* (On an irritated dog) 326; 349–351 (On swine).

434 *Ibidem*, 326. *Si caballus cum pede, si boves cum corno, si porcus, cum dentem hominem intrigaverit aut si canis morderit, excepto ut supra, si rabioso fuerit: ipse conponat homicidium aut damnum cuius animales fierit, cessante faida id est inimicitia, quia res muta fecit, nam non hominis studium.*

'dumb things', while on the other – people as a species. The feature differentiating the said animals from people was here the ability to talk, which was treated in this context as a manifestation of intelligence, and in particular the conscious planning of a crime. Let us draw attention here to the fact that this regulation concerned an incident in which an animal appears in the role of 'perpetrator'. For our research more important are the regulations which talk about animals as the victims of crimes. For thanks to these it will be possible to compare them with the analogical regulations on people.

So let us move to the question of the differences in the way the body of animals and people was perceived, and in particular the body of slaves. *Edictum Rothari* contains a group of regulations on crimes involving the damage to certain parts of a horse's body. These are titles 337 (*De caballo plagato*), 338 (*De coda caballi*) and 339 (without a heading).[435] The first mentioned concerns a situation where someone has cut off the ear or knocked out the eye or brought about some kind of damage to the body of someone else's horse. In this case the perpetrator received the wounded animal so that he may give its owner compensation for the loss by giving him back a similar horse (*ferquido*). Title 338 (On the tail of a horse) speaks in turn about the cutting off of the tail, and strictly speaking the hair. In this case, differently than in the previous regulation, the perpetrator was to pay 6 *solidi* in compensation.

Title 339 concerns striking or wounding, as well as causing other forms of harm to someone else's horse. In this case the perpetrator was obliged to treat the injured animal and to provide the owner of the injured animal his own steed until the end of the treatment period. If the animal returned to health it was to go back to its owner, though if it were to die as a result of the injuries or wounds inflicted then the perpetrator was to give the owner another, similar horse. As can be seen the content of this regulation coincides partly with the regulations from title 337, in relation to which there was possibly a more detailed development.

We shall now look at the solutions adopted in these three regulations. They show both certain similarities as well as distinct differences in relation to those rules on damage to the human body. The common feature is the treatment of the animal organism (in this case that of a horse) as the object of crime. A manifestation of this is the precise description of the damage to the individual parts of

---

435 ERot, 337. *De caballo plagato. Si quis caballum alienum aurem aut oculum excusserit aut aliquam laesionem corporis fecerit, recepiat ipsum, qui laesus factus est, et reddat ei ferquido, id est similem*; 338. *De coda caballi. Si quis caballum alienum coda cappellaverit, id est setas tantum, conponat solidos sex*; 339. *Si quis caballum alienum plagaverit aut aliquam laesionem fecerit, tunc dominus illius cavalli retradat cavallum ipsum illi homini, qui ei laesionem fecit, ut ipse eum sanet* [...].

the body such as the eye, ears, or tail. Besides, attention is also drawn here by the formulation *aliqua laesio corporis*, which was used in another of the *Edictum*'s regulations in relation to man.[436] In the case of cutting off the tail the perpetrator had to pay the animal's owner compensation. This solution most closely mirrors the regulations applied to slaves (the owner received compensation in the case of damage to the body of his slave). One may add, however, that the designation of a specified amount of compensation for a specified type of damage and injury was applied in relation to all Lombard social groups. This meant that in relation to an animal's body and more precisely its parts, in the case of damage, there was applied the self same regulation as with a man's body.

There is, however, a fairly significant difference. In the case of serious damage, within which as one can see the knocking out of an eye and the severing of an ear were included, as well as other related unspecified woundings or other disturbances of the organism, the compensation for loss involved the replacement of the victim's animal by another - one without defects. Compensation for this type of damage had, therefore, a specific character. The damage could not be regulated on the basis of 'payment' for the bodily harm. Such a means for the resolving of disputes between perpetrator and the victim's owner was not applied in *Edictum* for any other category of Lombard society. Even in the case of slaves that were after all equally the property of their masters in a similar way to animals, there was not applied the principle of exchanging a damaged 'item' with an undamaged one. The adoption of such a regulation, even though it would have been presumably favourable for the owners themselves was, however, not implemented for some reason in relation to people. The treatment of a horse (or other animals) as an 'object' subject to exchange is in accordance with the cited above definition of animals as 'dumb things'.

The difference between the body of a man and the body of an animal as sketched in the light of the discussed titles from *Edictum Rothari* is fairly vague. Even the similarity in the course of action taken in relation to slaves and horses is only partial. It appears that this example illustrates a wider phenomenon, one visible in other barbarian codes, and namely the transfer of certain principles applied in relation to the human body to domesticated animals, constituting (in a similar way to slaves) someone's property. Certain similarities concerning the treatment of human and animal bodies resulted presumably from this way of understanding on the part of the creators of the *leges*. In title XIV (On animal damages and their compensation) of Bavarian law there is talk of: the knocking out of one eye of a horse, bull or other quadruped (compensation: 1/3 of the value), the cutting off (1/3 of a *solidus*) or damaging a bull's horn (2 *saicas*) and the

---

[436] *ERot*, 27.

cutting off of a cow's horn (2 *saicas*), the cutting off of the tail or ear of a horse known as a *marach* (1 *solidus*), a medium sized horse known as a *wilz* (half a *solidus*), a smaller horse, an *angargnago*, (1/3 of a *solidus*), the cutting off of the tail or ear of a bull (1/3 of a *solidus*) or a cow (2 *saicas*).[437] We can find similar regulations also in *Lex Alamannorum*. There talk is on the knocking out of an eye or cutting off of the tail of a horse of the *marach* type (compensation – 3 *solidi*), a medium sized horse (1 ½ *solidi*), a beast (*iumentum*) – 1/2 *solidi*.[438]

In these regulations not only is attention drawn by the use of terms such as amputation or other means of removing particular parts of the body (the eye, ear tail, horn) but also the specific classification of the various types of animal (horse, bull, cow, other quadruped or beast), and also their dimensions and type (horse of the type: *marach*, medium, inferior). The compensation herein was varied depending on the categories of animal devised according to the criterion of species and varieties as well as the type of damaged part. This system, it should be emphasised, recalls in relation to its structural dimension the practices applied with regard to people. The division of horse into three categories is recollective of the three categories of freemen appearing, for example, in Alamann law. There was applied in Bavarian law the principle whereby compensation in the case of concrete bodily damage was to constitute a defined part (1/3) of the animal's value (horse, bull, other quadruped). We know of this practice from *Edictum Rothari* where it concerned both freemen and slaves.

We should incorporate into our investigations also those regulations in which the border between the world of people and animals is somewhat differently drawn. Particularly interesting here is the situation of slaves. We come across such a case in *Edictum Rothari*. The matter concerns regulations on bringing about miscarriages in cattle, mares and slave women as well as their death as a result. These three titles (332. *De vacca praegnante*, 333. *De equa praegnante* and 334. *De ancilla praegnante*)[439] were placed in the group of regu-

---

[437] *LBaiw*, XIV, 8–14; cf. also section XX (On dogs and their compensation), 1–9, concerns the killing or theft of 9 categories of dog. Their classification was conducted on the basis of the tasks they performed; these were various types of hunting dog (tracking, hounds, dachshunds etc.), shepherd dogs and guard dogs. In this sense the list somewhat recalls the regulations on the killing of slaves of various specialists, particularly in *Edictum Rothari*, with one significant difference – in 8 out of 9 cases beside the payment of money the offender was to return or provide in exchange a similar dog.

[438] *LAl*, LXXII, 1–3.

[439] *ERot*, 332. *De vacca pregnante. Si quis percusserit vacca preganantem et avortum fecerit, coponat tremisse uno. Et si mortua fuerit, reddat eam, qualiter adpraetiata fuerit, simul et pecus*; 333. *De equa pregnante. Si quis percusserit equa pregante et avortum fecerit, conponat solidum unum; et si mortua fuerit, reddat eam, simul et pecus, ut su-*

lations on animals (titles 309–349), however even the very arrangement of cattle, mares and slave women points to the Lombard lawmakers placing the latter within the category of live goods.

These regulations were edited in accordance with an almost uniform scheme: in the case of the death of the foetus, resulting from a blow by the perpetrator to the mother, compensation was set (a calf - 1/3 of a *solidus*, a foal – 1 *solidus*, a child – 3 *solidi*). If, as a result of the miscarriage, the mother also died, in the case of cattle and mares the perpetrator was to give back as much as the value of the mother and foetus. The words used in the first two regulations appear important: *reddat eam, qualiter adpraetiata fuerit, simul et pecus*. In the title on slave women the perpetrator was to pay her compensation. Yet the wording is different: *conponat eam, simul et quod in utero eius mortuum est* (i.e.,: he/she pays her compensation and for that which is dead in her womb). Therefore certain differences between the discussed regulations are visible. A slave women had in the case of death a fixed, standardised amount of compensation written into the legal code.[440] If a cow or mare was killed the compensation was every time individually determined for the concrete animal.[441] The rigidly legal-

---

pra; 334. *De ancilla pregnante. Si quis percusserit ancilla gravida et avortum fecerit, coponat solidos tres. Si autem ex ipsa percussura mortua fuerit, conponat eam, simul et quo in utero eius mortuum est.*

440 The value of compensation for a slave woman was regulated by title 127, on the effects of bodily damage (*Omnes vero plagas aut feritas tam de aldius quam de servus ministeriales, seu servi atque aldias aut ancillas, que inter eos eveniunt* [...]). In the case where a slave woman (or a person belonging to another of the mentioned categories) died as a result of wounds within the course of one year from their infliction, her master was to be paid compensation ([...] *si autem de ipsas plagas mortuus fuerit intra anni spatium sicut subter adnexum est, ita domino componatur*) at a rate set in titles 130–136, as appropriate to the category of slave woman, i.e., from 16 to 50 *solidi*, including the amount already paid by the perpetrator for bodily damage ([...] *et quod pro plaga acceptum est, in ipsa summa compositionis mortui reputetur*). Attention is here drawn by the formulation 'the sum of the deceased's compensation', which referred also to slave women.

441 We shall recall here title 339 discussed above on damage to the body of a horse and its results (*Si quis caballum alienum pagaverit aut aliquam lesionem fecerit* [...]); in a case where as a result of injuries the said horse died, the perpetrator had to supply in exchange a horse of the same type ([...] *si autem ex ipsa lesion mortuus fuerit, reddat alium similem*). There is no mention here as to the 'amount of compensation' for the dead horse as it was not established; the same applies to mares and other domesticated animals. In the case of stealing dogs (title 329) the perpetrator was to return nine-times its value (*Si canem furaverit, sibi nunum reddat*). Therefore, one can see that titles 332 and 333 were no exceptions. In all of these regulations there is in force the principle of each time evaluating the worth of the given article and equating the loss either in money or in

ly set amount of compensation for the death of an unborn child was not ordinary compensation by a mark of status. For this was only applicable to people. In addition in title 334 there is mention of the womb of a slave woman, something that is absent in the case of cows and mares. The fact that in this very place there occurs a specific bodily object (*uterus*), which also occurs in animals, is not accidental. It fulfils in this context the role of a feature differentiating the slave woman from the mentioned females.

In *Edictum* there exists also a clause (title 75. On a child if it was killed in its mother's womb) relating to the analogical act committed in relation to a freewoman.[442] In the case whereby the unborn child died and the woman lived, the perpetrator was to pay compensation that constituted a half of the *wergeld* value for the mother as established 'according to her nobility'. If the mother died – he paid her *wergeld* and compensation for the child, i.e., a half of her *wergeld*. The difference between title 75 and the three regulations discussed above was therefore dependant on that fact that in the case of a free woman the compensation for the death of the unborn child was a derivative of the mother's *wergeld*, while for a cow, mare and slave woman it was established as if independently. For first and foremost it possesses a legal status resulting from the mother's birth status, something that cannot be said about a calf, foal, or the child of a slave woman.

When the matter concerned the death of a woman things were different. Both a freewoman as equally a slave woman had legal status, thought obviously different. This is borne out by the use of extremely similar terminology. For in both titles (75 and 334) there appears in relation to women the formula *componat eam*. Besides, in title 75 there is used the formulation *quod in utero eius mortuum fuerit*, which also described the unborn child of a dead slave woman. Therefore, it seems that despite the indicated differences, the wombs of both women were treated in a similar way. Therefore once again it turns out that the said 'bodily object' (in a similar way to the collection of body parts) was viewed as a common feature for a freewoman and slave women. A person's body was in *Edictum Rothari* an element of legal status for the representatives of all social groups, therefore one may assume that it constituted a human trait and sign of

---

the form of a living replacement of the self same value. Attention is also drawn by the certain similarity between titles 127 and 339; these were modelled on the regulation on the effects of damage to the body of freemen (title 74). We are dealing, therefore, with the transfer of certain resolutions from title 74 to *aldii*, slaves and horses.

442 *ERot*, 75. *De infante, si in utero matris occisus fuerit. Si infans in utero matris suae nolendo occisus fuerit ab aliquem: si ipsa mulier libera est et evaserit, adpraetietur ut libera secundum nobilitatem suam, et medietatem, quod ipsa valuerit, infans ipse componatur. Nam si mortua fuerit, componat eam secundum generositatem suam, excepto quod in utero eius mortuum fuerit, ut supra, cessante faida, eo quod nolendo fecit.*

affiliation to society. From comparative analysis it equally results, however, that a slave woman's unborn child was treated in a similar way to the foetuses of cows and horses.[443]

In talking about the body of a person and an animal as objects of legal regulation it follows to address equally the question of the human corpse and an animal's dead body. The legal resolutions on the stealing of human bodies and graves has been addressed above. It follows only to mention that in certain *leges* there existed regulations on crimes of this type both for freemen and slaves. The corpses and their burial was therefore a common feature of both these groups. We shall see in turn how in the oldest Lombard legal code the body of a dead man or a dead animal was perceived. Title 335 (*De animale excoriato*) talks about a case whereby a wolf killed somebody's animal (domesticated) while another man illegally skinned it and hid; title 336 covered the skinning of a dead animal which had been found in a river or other place.[444] We are here dealing with the utilization of a dead animal's body, which was a crime as it constituted the depletion of someone's property and therefore a form of theft. The body of a dead animal was therefore an object which could be used in various ways (in a similar way to someone else's meadow or farm tools).

In Bavarian law, as we wrote in Chapter II, there existed a regulation that talked of the damaging of unburied corpses. This crime was, however, of a different character – this was desecration. Presumably it is not a question of coincidence that there are no regulations in the barbarian *leges* which deal with gaining benefit from a human corpse (either free or slave), although technically this would have been possible. The removal of skin from a dead animal did not constitute a violation of its honour or dignity; it was though the gaining of material benefit to the detriment of the owner. The bodies of dead animals had only material value; while the corpses of men had first and foremost honour. Hence there arose the legal protection of graves and the bodies of the dead therein buried, while on the other hand the lack of regulations in relation to animals.

---

443 We shall add that in the case of the accidental death of the small child (*infans parvus*) of a slave known as *massarius* (a farmer) the perpetrator was to pay the compensation set by the judge proportionate to the child's age or the profit it could have brought; cf. *ERot*, 137. *Si infans parvus de massario occisus fuerit. Si quis infantem parvulum de servo massario casu facientem occiderit, arbitretur a iudice; secundum qualem aetatem habuit aut qualem lucrum facere potuit, ita componatur.*

444 *ERot*, 335. *De animale excoriato. Si lupus animalem cuiuscumque occiderit et aliqui eum nescientem domino excoriaverit et celeraverit et per proditurem inventum fueri, componat solidis duodecim;* 336. *Si in flumine animal mortuus fuerit aut ubicumque, et ab alio homine, cuius non fuerit, excoriatus fuerit: qui eum excoriaverit, componat solidos duodecim.*

Summing up the above analysis we may formulate one more conclusion: the creators of the legal codes perceived a man's body not only as an object of law, but also a moral object. For crimes against the body violated not only its integrality and inviolability but also the dignity of the person both in life as in death. This aspect of human corporality is extremely visible in the barbarian *leges* when we compare the regulations on the body of man and animal. On their basis we may state that dignity was to some extent written into a person's corporality. Therefore the presence of this moral aspect in the researched regulations enhances the thesis that the body was a factor linking all the social groups appearing within the various legal codes, it was their common feature, separating the human world from that of the animal.

Conclusion

The problem area from which we commenced our study into the human body as an object for legal regulation as contained in the barbarian *leges*, is that of the meaning and application of the word *corpus* as well as its Germanic equivalent terms *hreu, hreo, hraiwa*. An analysis of the content of the individual codes has shown that in the majority of them these words refer to the body of a dead person – a corpse and not to that of a living man. The concept of body thus understood appears first and foremost within the context of defiling corpses. In some codes the word *corpus* meant the trunk or torso while the Old Germanic words cited above meant 'abdominal cavity'. Words representing the body of a living person do not appear in the oldest Anglo-Saxon code, which was the only one in the set researched that was written in its indigenous language and not in Latin. Yet it is known that there existed within the Anglo-Saxon language the words *lîc, lîchoma*, which in other sources (e.g., in *Beowulf*), were used in this meaning. We can see another practice in Visigoth laws where the word *corpus* is used in relation to the body of a living man. We can observe the same in the oldest compilation of Lombard laws. Visigoth law contains, however, also the term *pars corporis* which relates to a living body. This points to an understanding of the body as a holistic structure composed of parts. In a part of these there appears the designation *membrum* (member), which was used, however, not in the meaning of 'a part of the body' but as a designation for a concrete individual 'object' (for example, the innards).

In the majority of the *leges* individual bodily 'objects' such as the hand, eye or thumb, which often appear in them, were connected with the concept of man, and not with an autonomously understood category of 'the body'. An injured hand was understood as something that belonged to an injured person. A man was not perceived as an abstract bodily structure composed of parts but as a 'living soul', or a 'living entity'. A person thus understood as a concept replaced to some extent the concept of the body. Crimes involving the use of violence, which constituted the subject area of our research into legal regulations, concerned, in the conviction of those who penned the rules, concrete individual elements of a man's bodily shell. The use of the concept of the body of a living person understood as an entirety composed of parts is, in the laws of the Visigoths and Lombards – it seems – a manifestation of the influence of Latin written culture on the editing process of law regulations. In other codifications where instead of the concept of the body there is used the concept of a man, we are dealing with the dominating influence of a mindset formed by the culture of the spoken word.

The absence of words meaning the body of a living person in the majority of legal codes examined is surprising for the modern reader. For the specific feature of the barbarian *leges* are the more or less elaborate sets of regulations on damage and violation of particular parts or fragments of the body. We are dealing, therefore, not so much with the body as with collections or sets of individual parts of the body. In the majority of the said codifications the given sets had a specified arrangement, the model for which was the figure of man as seen from head to foot. In such an order the regulations on crimes against the integrality of the body and the bodily inviolability of man were written down. The creators of the examined *leges* did not inform one, however, about the use of the human form as a model which organised the structure of the groups of regulations that interest us. This model operated as if beyond the text. It follows to add, however, that there exists a group of *leges* in which the regulations on crimes against the body were not enumerated in accordance with the scheme of the human form.

An important element in our considerations is the investigation of the reasons for the use of the word *corpus* in the majority of the *leges barbarorum* in the meaning of 'corpse' as well as the lack of reference to the body of a live man. Explaining this phenomenon may be linked to three factors that moulded the content of the Germanic legal codes. The first of these is the subject of interest and the normative functions of the *leges*. The subject under discussion is in the case of the researched barbarian legal regulations physical violence, and in particular the violation and damage of particular parts of the body. This subject resulted in a fragmentary perception of a man's body. The function of the law involved, in this case, the regulating on situations that had arisen as a result of the undertaking of certain acts. The concept of a living body did not appear in them, for the creators of the regulations were interested in cases of damage to concrete parts of the body. The category of the body was therefore of no need to the codifiers. All the same the said regulatory function of the law resulted in the creation of an extensive list of its individual components.

The second factor which may be taken into consideration as a reason for the said specific understanding of 'the body' is the way in which knowledge is created and transferred within a culture of the spoken word. The absence of the concept of 'body' understood as a complex autonomic structure in relation to the concept of man resulted, in the majority of Germanic legal codes, from the oral mentality of those who drew up these regulations and the way in which ideas connected with it were expressed. This is visible in the laws of the Franks, the Anglo-Saxons, the Alamanns, the Bavarians, the Saxons, the Thuringians and the Frisians. The concept of the body was replaced in them by the concept of man. There is also an absence of categories of 'body parts'. The individual members were not there categorised as either the body or its parts. They constituted a set of elements unconnected with each other.

The Visigoth and Lombard legal codes, which to a large degree were subject to the influences of Roman written culture, contained more abstract concepts of the body. It follows, however, to note that even in these codifications the regulations on crimes against the body were created separately for its individual parts in such a way that a subsequent violation was treated as a separate case. In narrative texts from the early medieval period (for example, the *Histories* of Gregory of Tours, Einhard's *Life of Charlemagne*), ones created via the culture of the written word, we are dealing with defined and abstract concepts of the body. It had a holistic character – it was a complex system of mutually connected parts (members). The body was perceived separately than in relation to man himself.

The third way of explaining our problem refers to the religious factor of 'world outlook' that shaped the texts of the *leges*. In the case of the body the best effect of this factor is to be seen in relation to its posthumous state. We are here dealing with a way of thinking about the body of a deceased person in the categories of traditional Germanic (pagan) beliefs. This is pointed to by the regulations contained within *Lex Baiwariorum*. There is talk about the mutilating of unburied bodies. The nature of the said damage recalls that inflicted on living persons. The reason for the creation of regulations on the mutilation of corpses was the codifiers' conviction that posthumous damage was equally damaging for the victim. Therefore, it seems that in the barbarian *leges* we are not dealing with a Christian concept of man as a spiritual-bodily being of a dichotomous structure whether in the case of the corpse or the body of a live person. This concept arose from a 'world view' whose source was the early medieval beliefs of the Germanic peoples superficially modified by the Christian 'anthropology' of the time. The death of a man was a situation which 'revealed' and emphasised the significance of his bodily shell. Hence the use of the word *corpus* in the *leges* in the meaning of a corpse.

Another key problem for the subject undertaken in this study is that of man's body as the object of criminality. For in the barbarian *leges* a man's body appears first and foremost in the regulations on various forms of damage to the body substance or violations of its inviolability. The research conducted has shown significant differences in the number and type of body parts and fragments which have been incorporated in the thirteen analysed codifications. Each of them possess their own features in relation to the selection of damaged body parts. A comparison of the said collections proves that in the individual codes the lists of body elements could have been expanded without limitations through the addition of new parts, the setting apart of their smaller fragments (e.g., parts of the finger) or the listing of subcategories (e.g., various types of teeth).

Besides, a comparison of the lists of body parts appearing in the particular codes has resulted in the establishment of a catalogue of the said bodily 'objects' on the basis of the contents of all the codes. This covers 70 items. The collection may be divided according to body 'zones' (the head, trunk, upper limbs and lower limbs). Research into the frequency with which the particular elements from this list appear in the whole *leges* set shows that only 15 body parts may be considered to have been universally taken into account by the compilers of the *leges*. These were: the eye, nose, ear, skull, trunk, penis, thumb, index finger, middle finger, ring finger, little finger, hand, the whole arm, the foot, and undefined places on the skin.

The next subject of our investigation are the types of crimes against the body which were mentioned in the researched codifications. Generally these acts may be divided into those which involved damage (permanent or non-permanent) to the body substance as well as those which constituted violation of bodily inviolability. In the source research material there appear a dozen or so variations of crimes of this type. They differ in the character and degree of severity (both physical as moral). One may enumerate here: amputation, acts aimed at removing the organs of perception, castration, causing paralysis, bone breaking (including damage to the skull), piercing (including complete piercing), disfigurement (particularly the face), bloody woundings, blows, slapping in the face, cutting off hair or stubble, pulling hair. Such an extensive set of crimes does not appear in any of the researched codes. Each of the codifications contained a somewhat different selection of acts of this type. Comparative analysis of the appearance of the enumerated types of damage and violations has shown that the differences between the researched codifications are in this way quite small (incomparably smaller than in the case of the number of body parts). The crimes here listed concerned first and foremost a man's body, although in certain legal codes the regulations containing descriptions of acts of this type were *en bloc* applied to a woman's body. Besides, in a part of the barbarian codes there existed regulations that talked exclusively about crimes against the female body. These were first and foremost acts of sexual harassment (e.g., indecent exposure and grabbing the vulva).

In the overwhelming majority of cases the organisational structure of the wording on crimes was the listing of body parts and a cataloguing of the various forms of bodily damage and violation. In other words, the specific types of crime were classified within the list of body parts and not the other way round. The order in which the said acts were enumerated was strictly connected with the vision of the human body adopted by the codifiers. As we have earlier ascertained they understood the body in terms of a man's figure or shape in the literal meaning of the word. Such a perception of the human body determined the order in the description of crimes committed against it. Nothing equally points to the appearance of some type of crime automatically resulting in the consideration of a given body part.

An important problem in research into the regulations on the categories of crime here of interest to us is the role fulfilled by the descriptions of the said acts as a textual mechanism creating an account of the body or more precisely its components. The specifics of this type of description depends in the majority of cases on narrowing them down to a definite point or body fragment i.e., that which was the object of the crime. However, in certain (few) regulations there is equally talk about parts adjacent to those that underwent damage. The majority of the discussed regulations contained a laconic description of a specific type of damage to a given body part. In a small part of the descriptions we come across information about permanent effects brought about by certain crimes. These references significantly increase the scope in which these regulations present the victim's body. The consequences of body damage were varied in character: one of their groupings are cases of disabling the motor functions of the limbs (e.g., the inability to bend the arm at the elbow or hold objects); a specific result of this type of damage was the causing of impotence; other forms of body damage involved impairment to the organs of perception (e.g., an inability to speak); another grouping was disfigurement of the face (e.g., the cutting off of the lower lip making the person unable to retain saliva i.e., constant dribbling).

The significance of the descriptions of the said permanent effects for the examined crimes also involves the fact that we are dealing with a shift of emphasis from the damaged part of the body to its function in the organism (e.g., talk is not simply of the ear but of the ability to hear) as well as its appearance (e.g., deformation of the ear). Such descriptions prove that the creators of the *leges* were interested not only in the substance of the body but also man's corporality. The said corporality is visible also in the regulations that speak of exposing a woman's body. In this case of significance were not only the parts of the body (head, legs, vulva) but the very fact of the body's nakedness.

The problem of crimes involving the disfiguring or violation of specific parts or fragments of the body is connected, in an inseparable way, with the permanent sums of compensation to be paid to the victim (or his/her master) by the crime's perpetrator, sums determined by the codifiers. The subject of our considerations was an attempt to establish the mechanisms which resulted in the ascribing of specifically determined sums for specific types of damage to particular parts of the body. The starting point are two hypotheses on the reasons for the said compensation rates. The first states that the amount of compensation seen in the individual *leges*, represented compensation for the forms of violation and damage there outlined. The second hypothesis says that the amount of compensation represented a form of ransom paid by the perpetrator to the victim for renunciation of *faida*, that is resignation from revenge for the injury caused. Neither of these adequately explain the problem of the causes and mechanisms of compensation allocation.

Research into the payment tariffs contained in the individual barbarian codes show that the determining of their value was not based on an estimation of the rate of compensation for each individual case (the type of crime and the body part) separately but on their allocation to an already determined rate characterised by 'thresholds'. In each of the codes examined we are dealing with a arrangement of compensation rates developed to a lesser or greater degree. In many cases there is a visible numerical dependence amongst the amount of subsequent rates. In the particular codifications they constitute a multiple of the figure or figures selected by those compiling the given legal code. Often the highest amounts of compensation also show a clear correlation with the value of payment to be made in the event of killing a man (*wergeld*).

Analyses of the said tariffs also point to the fact that the amounts of compensation allocated in the particular *leges* created comprehensive systems. It seems grounded to presume that in the case of subsequent crimes payments were not set in an evolutionary way but once for the whole set of damages and violations. The value of the said sums was dependant on several factors. These included: the role of the given body part in the human organism (different in the case of the eye, nose, arm or finger); the severity of the given type of damage (different in the case of, for example, severance and wounding), its influence on the state of health and movement ability as well as the appearance of the body. Objective conditions of this nature, although undoubtedly important, were still subordinated to the threshold system of compensation.

The existence of a hierarchy of body parts resulting from the role they played in the functioning of the human organism had a limited influence on the determining of the amount of damages. A comparison of the payments set in the case of the self same crimes in particular codes shows that there did not exist a single universal value hierarchy for the parts of the human body. The numerical proportions between the payments for damaging certain body fragments (e.g., the hand, thumb and big toe) which appear in the individual barbarian codifications, differ amongst themselves. Besides, the same rate of compensation was in force in the case of different types of crime and different parts of the body. These assertions lead one to the conclusion that the amounts of compensation set in the legal codes under consideration did not reflect the objective value of bodily loss resulting from the damage to the given body part. One may only state, on the basis of the compensation tariffs, that their creators considered certain parts of the body to be more valuable than others, as well as certain types of damage to be more or less severe. The valorisation of the body had, therefore, a relative character. The numerical dependence between the *wergeld* value and the certain compensation rates that is visible in many of the *leges* cannot be interpreted as explanation for the criteria adopted in setting payment rates. For we do not know what the basis was for the establishment of *wergeld* payments. The rate of compensation expressed therefore the social evaluation of the severity of particular crimes.

An important motif of the research is the appearance in a part of the barbarian legal codes of the phenomenon of allocating different rates of payment for the same bodily crime and murder depending on the legal and social status of the parties involved. The first question that needs explanation are the possible reasons for social differentiation in compensation fees. Various hypotheses explaining this phenomenon are discussed: the actual physical differences amongst the various status groups; the conviction that certain bodily traits (particularly that of beauty) were somehow allocated to particular 'orders' (the nobles, freemen and slaves); the relations of forces amongst social groups; the belief on the part of the creators of the laws as to the inherent inequality amongst the social groupings belonging to a given people, which expressed itself in their separate legal status. Consideration of these leads one to the conclusion that the most likely way of explaining the said differences is through the category of legal status.

We have devoted a lot of attention to one of the characteristic features of the barbarian *leges* – the specific connection between notions on the structure of tribal society and the regulations on crimes against the body. The first problem connected with this is the model depiction of the social composition of a given tribe appearing in the particular codes. This question has been presented on the example of the law regulations of the Frisians, the Lombard *Edictum Rothari* as well as the laws of the Salian Franks. In Frisian law the 'image' of society is contained first and foremost in the regulations on the killing of the nobles, freemen, half freemen, and slaves (title I) as well as in those regulations that speak of crimes against the body where there is mention of the noblemen, freemen and half freemen. Each of these categories had a separate *wergeld* and damages rate. These regulations referred to the male part of the tribe. Only two paragraphs (out of 89) of title XXII (On wounds) talks of crimes against the body of a woman. Women appear first and foremost in those regulations on sexual crimes. There appear here: noblewomen, freewomen, half-freewomen and slaves. On the basis of the amounts of compensation given in title XXII one may state that the body of a freeman constitutes the model which was to serve in relation to both noblemen and half-freemen.

In the Lombard *Edictum Rothari* the model 'image' of social structure is visible first and foremost in the regulations on crimes against the body (titles 42-127). In turn, reference is made to three groups: freemen (only men), half-freeman (*aldii*) and qualified slaves as well as rural slaves (both categories – men and women). There appears directly after these regulations a set of rules on murder and manslaughter, in which there is mention of half-freemen together with 9 categories of slave (titles 129-136). In *Edictum Rothari* there is no mention as to the *wergeld* of freemen. There is talk, however, of payments for the killing of a freewoman. There also exist regulations for the raping of half-freewomen, free women and slave women.

A third analysis deals with the laws of the Salian Franks (*Pactus legis Salicae* and *Lex Salica*). In this codification the set of regulations on crimes against the body concerns only a single social group – that of freemen (titles XVII and XXIX). A conceptualisation as to the structure of Frankish society may be ascertained in the regulations on murder. This 'picture' is, however, fairly hazy as the subsequent titles in which there is talk of murder contain contradictory data. In *Pactus legis Salicae* there are listed 16 different categories of murder. The differences in social position was only one and not the most important criterion in their differentiation. These occur, first and foremost, amongst men, where next to ordinary freemen are to be found dignitaries (*grafion, antruscion, sacebaron*). A large group of these concern women, who are divided into subgroups depending on their ability to give birth. Here also are children up until the age of 12 as well as unborn children of both sexes. Mention is also made of slave men and slave women.

A comparison of the Frisian, Lombard, and Frankish regulations shows that we are dealing with three separate ways of shaping the model image of tribal social structure. The image of the Frisian community in the light of title I (*De homicidiis*) was highly homocentric, i.e., narrowed down to the male part of the population, though also comprehensive in covering all estate groups (the nobles, the ordinary freemen, half-freemen and slaves). The 'image' of the Frankish community as we may observe in *Pactus legis Salicae* and *Lex Salica* was dominated by two groups – namely women and children. Against this background men – ordinary freemen – occupied a less prominent position both with regard to the value of the *wergeld* as the numerous subcategories in the male population. The regulations of *Edictum Rothari* combine some features of the two presented versions. On the one hand, it possesses a comprehensive character given its schematism – covering freemen, half freemen and slaves. While on the other, there are incorporated within it, although not totally, women (half freewomen and slave women). The guiding idea in shaping the said classifications of social groups within the Germanic barbarian codes was the model concept of social order, though there was no intention of depicting with it all groupings and categories of person.

The next important problem is the relations between the textual presentations of man and the visions of social order that we many observe in the legal codes of the particular Germanic peoples. Here the matter concerns particularly the means in which the said presentation of the human body was entered into the classification of the social groups adopted in the given codification. This problem is discussed on the basis of the example of the oldest legal code of Lombard law, the law of the Bavarians as well as of the Ripuarian Franks. The sets of body parts of a freeman's body constituted in *Edictum Rothari* (as equally in other Germanic codes) its own form of matrix, which was copied in relation to other status categories (those of freewomen, freed men and slaves). A free (common) man as a representative of his category was, in the light of the regulations under discussion, his own centre as well as the axis of tribal society. His body was therefore the model for the human body in general.

The Lombard legislators treated the body as the factor which assimilated or even placed on the same plane the representatives of the three social groups already mentioned: freemen, half freemen and slaves. Another argument that the body was in the conviction of the creators of barbarian law a common feature for individuals belonging to different legal-social groups, including equally slaves, are the regulations on the defilement of corpses and graves. In the law of the Alamanns, Salian Franks and Bavarians we find regulations which protected the graves and corpses of both freemen and slaves.

The practice of treating a human body as a feature common to all men that belonged to the three main status groups (freemen, freed men and slaves) is visible also in *Lex Baiwariorum*. In the Bavarian codification (in a similar way to *Edictum Rothari*) there was in force the principle that the image of a freeman constituted the model which was to be copied in relation to freed men and slaves. There was in force equally the same order in the enumeration of body parts and crimes that involved their violation. A similar way of thinking about the social-legal dimension of man's physicality can be observed also, although to a somewhat lesser degree, in the codification of the Ripuarian Franks – *Lex Ribuaria*. The structure of the male part of this people seen through the prism of a system of damages for crimes against the body comprised two groups: freemen and slaves.

There are several different solutions presented to us in the individual codes as far as the relations between the model 'images' of social construction and the perception of the human body are concerned. The legal codes of the Salian Franks and the Lombards (first and foremost *Edictum Rothari*) constitute an example of the different ways of designating the links between the model presentations of the body and society. In *Pactus legis Salicae* the image of tribal structure is contained within the regulations stipulating the amounts of compensation to be paid for the killing of representatives of various social groups. In *Edictum Rothari* the model presentation of social structure is contained within the regulations on the bodily violation of the three main social groups. The difference between the two models (the Salian and Lombard) is twofold in nature. Firstly, in the codifications of the Salian Franks the textual presentation of the human body does not play any role in the account of the social structure scheme, while in the Lombard it was to have, in this respect, a key significance. Secondly, which is connected with the first, in Salian law a man's body as a comprehensive structure was reduced to the body of a freeman, while in Lombard law this image was extended to the remaining social groups (although with the exclusion of freewomen).

Another problem which is connected to the above issue is the connection of the human body, as an object of crime, with the hierarchical social order. The most visible and really significant manifestation of the relations between man's body and social order that emerges from the regulations of the barbarian *leges* herein investigated is that every, even the most minute, 'bodily object' (for example, a single tooth, a part of a finger, a toe) was treated in them as an element of the said order. This happened so because each part of the body was evaluated from the viewpoint of the legal and social status of the individual. Differences in status represented therefore differences in the amount of compensation. Legal status was somehow written into the human body. An individual's set of body parts constituted an element of the idea of social order. Individual members or fragments of the body were consequently not merely legal objects but also their collection as a whole was incorporated into the system of the power relations (of authority and subordination, of hierarchy) that existed (in the light of the researched codes) amongst social groups.

The last question connected with the social dimension of the body is its role as a determinant of affiliation to the human world. The main subject of investigation is the comparison of regulations for the human body and that of animals. Here the matter in particular concerns the legal situation of slaves and domesticated animals. Research into the regulations on animals has shown a certain difference in the treatment of the animal body and those regulations governing people. This is the principle that in the case of severe damage to the body of an animal the perpetrator was obliged to compensate its owner by giving him an undamaged animal of the same value. At the same time it can be seen, however, that this phenomenon involves the transfer of certain principles used in relation to the human body to domesticated animals which (in a similar way to slaves) were somebody's property. The compensation set in the case of injury to domesticated animals (horses, bulls, cows, other quadrupeds or cattle) was various depending on their category, created on the basis of the criterion of species and varieties as well as the type of bodily damage. This system, it follows to emphasise, recalls in its structural aspect the practices employed in relation to people. The transfer of the legal solutions used in relation to people to livestock did not result in a blurring of the boundary between people and animals. This can be seen on the example of the regulations on human corpses and carrion. The body of a dead animal was an object that could be used in various ways (in a similar way to someone else's meadow or household equipment). The subject of the regulations governing these cases was the property protection of the owner's dead or injured animal. In relation to people the law protected the dignity of the deceased's corpse.

Abbreviations used in the footnotes and bibliography

Add. Sap. – Additio Sapientum [in:] Lex Frisionum
ADHE – Annuaria de Historia del Derecho Espagñol
Æthel – Æthelberth
Cap. leg. sal. add. – Capitula legi salicae addita
ERot – Edictum Rothari
HRG – Handwörterbuch zur deutschen Rechtsgeschichte
LAl – Lex Alamannorum
LBaiw – Lex Baiwariorum
LConst – Liber Constitutionum [in:] Leges Burgundionum
LFris – Lex Frisionum
LRib – Lex Ribuaria
LNG – Leges Nationum Germanicarum
LSal – Lex Salica
LSax – Lex Saxonum [in:] Leges Saxonum et Lex Thuringorum
LThur – Lex Thuringorum
LVisig – Lex Visigothorum
MGH – Monumenta Germaniae Historica
PLAl – Pactus legis Alamannorum
PLSal – Pactus legis Salicae
ZRG GA – Zeitschrift der Savigny-Stiftung für Rechtsgeschichte, Germanistische Abteilung

Bibliography

Sources

Breviarium Alaricianum. Römisch Recht im Frankischen Reich, ed. M. Conrat (Cohn), Leipzig 1903.
The Digest of Justinian, ed. Th. Mommsen, P. Krueger, A. Watson, vol. IV, Philadelphia 1985.
Die Gesetze der Anglesachsen, ed. F. Liebermann, vol. I, Aalen 1960.
Le leggi die Longobardi, ed. C. Azarra, S. Gasparri, Milano 1992.
Leges Alamannorum, MGH LNG, vol. V, 1, ed. K. Lehmann, K. A. Eckhardt, Hannover 1966.
Lex Alamannorum. Das Gesetz der Alemannen, Text – Übersetzung – Kommentar zum Faksimile aus der Wandalgarius-Handschrift Codex Sangallensis 731, ed. C. Schott, Augsburg 1993.
Lex Baiwariorum, MGH LNG, vol. V, 2, ed. E. von Schwind, Hannover 1926.
Leges Burgundionum, MGH LNG, vol. II, 1, ed. L. R. von Salis, Hannover 1892.
Leges Langobardorum, MHG Leges, vol. IV, ed. F. Bluhme, Hannover 1868.
Lex Frisionum, MGH Fontes iuris Germanici antiqui, ed. K. A. Eckhardt, A. Eckhardt, Hannover 1982.
Lex Ribuaria, MGH LNG, vol. III, 2, ed. F. Beyerle, R. Buchner, Hannover 1954.
Lex Salica, MGH LNG, vol. IV, 2, ed. K. A. Eckhardt, Hannover 1969.
Leges Saxonum et Lex Thuringorum, MGH Fontes iuris Germanici antiqui, ed. C. von Schwerin, Hannover 1918, pp. 51–75.
Leges Visigothorum, MGH LNG, vol. I, ed. K. Zeumer, Hannover 1902.
Pactus legis Salicae, MGH LNG, vol. IV, 1, ed. K. A. Eckhardt, Hannover 1962.
The Laws of the Earliest English Kings, ed. F. L. Attenborough, Cambridge 1922.
Einhardi Vita Caroli Magni, [in:] Fontes ad historiam regni Francorum aevi karolini illustrandam, pars I, ed. R. Rau, Berlin 1955.
Edda. Die Lieder des Codex Regius nebst vervandten Denkmäler, ed. G. Neckel, vol. I, *Text*, umgearb. Aufl. von H. Kuhn, Heidelberg 1962.
Gai Sollii Apollinaris Sidonii Epistula et Carmina, ed. Ch. Luetjohann, MGH SS vol. VIII, 1, Berlin 1887.
Gregorii episcopi Turonensis, *Historiarum Libri Decem*, ed. R. Buchner, vol. II, Berlin 1956.

Studies

Amira Karl von, *Die germanischen Todesstrafen*, München 1922.

Ausenda Giorgio, *Jural relations among the Saxons before and after christianization*, [in:] *The Continental Saxons. From the Migration Period to the Tenth Century: An Ethnographic Perspective*, ed. D. H. Green, F. Siegmund, San Marino 2003, pp. 113–131.

Azarra Claudio, *Introduzzione al testo*, [in:] *Le leggi die Longobardi*, ed. C. Azzara, S. Gasparri, Milano 1992, pp. XXV–XXXIX.

Bethurum Dorothy, *Stylistic features of the old English laws*, The Modern Language Review, vol. XXVII (1932), pp. 263–279.

Beyerle Franz, *Über Normtypen und erwieterungen der Lex Salica*, ZRG GA, vol. 40 (1924), pp. 216–261.

Bildhauer Bettina, *Medieval Blood*, Cardiff 2006.

Bognetti Gian P., *Frammenti di uno studio sulla compositione dell'Editto di Rothari*, [in:] *L'età longobarda IV*, Milano 1968, pp. 585–609.

Bynum Caroline W., *The female body and religious practice in the later Middle Ages*, [in:] *Fragments for a History of the Human Body*, ed. M. Feher, R. Naddaff, N. Tazi, New York 1989, vol. 1, pp. 161–219.

Bynum Caroline W., *Fragmentation and Redemption: Essays on Gender and the Human Body in Medieval Religion*, New York 1991.

Camille Michael, *The image and the self: unwriting late medieval bodies*, [in:] *Framing Medieval Bodies*, ed. S. Kay, M. Rubin, Manchester 1994, pp. 62–99.

Charles-Edwards Thomas M., *Law in the western kingdoms between the fifth and the seventh century*, [in:] *The Cambridge Ancient History*, vol. XIV, *Late Antiquity: Empire and Successors, A. D. 425–600*, ed. A. Cameron, B. Ward-Perkins, M. Whitby, Cambridge 2000, pp. 260–285.

Collins Roger, *Early Medieval Spain. Unity in Diversity, 400–1000*, London 1995 [1983].

Collins Roger, *Law and ethnic identity in the western kingdoms in the fifth and sixth centuries*, [in:] *Medieval Europeans: Studies in Ethnic Identity and National Perspectives in Medieval Europe*, ed. A. P. Smyth, Basingstoke 1998, pp. 1–23.

Collins Roger, *Visigothic Spain 409-711*, Oxford 2004.

Dilcher Gerhard, *Gesetzgebung als Rechtserneuerung. Eine Studie zum Selbstverständnis der mittelalterlichen Leges*, [in:] *Rechtsgeschichte als Kulturgeschichte. Festschrift für A. Erler zum 70. Geburtstag*, ed. H. J. Becker, G. Dilcher, G. Guadian, E. Kaufmann, W Sellert, Aalen 1976, pp. 13–35.

Fischer Drew Katherine, *Introduction*, [in:] *The Burgundian Code. Book of Constitutions or Law of Gundobad. Additional Enactments*, transl. K. Fischer Drew, Philadelphia 1972, pp. 1–10.
Fischer Drew Katherine, *Introduction*, [in:] *The Lombard Laws*, transl. K. Fischer Drew, Philadelphia 1973, pp. 1–37.
Fischer Drew Katherine, *Introduction*, [in:] *The Laws of Salian Franks*, transl. K. Fischer Drew, Philadelphia 1991, pp. 13–51.
Fischer Drew Katherine, *Law and Society in Early Medieval Europe*, London 1988.
Eckhardt Karl A., *Zur Enstehungziet der Lex Salica*, [in:] *Festschrift zur Feier der 200jährigen Bestehens d. Akademie der Wissenschaften in Göttingen*, Göttingen 1951, pp. 1–31.
Elsakkers Marianne, *Inflicting serious bodily harm: the Visigothic 'Antiquae' on violence and abortion*, Journal of Legal History, vol. 71/1 (2002/2003), pp. 55–63.
Elsakkers Marianne, *Abortion, poisoning, magic, and contraconception in Eckhardt's 'Pactus Legis Salicae'*, [in:] *Quod vulgo dicitur. Studien zum Altniederländischen*, ed. W. Pijnenburg, A Quak, T. Schoonheim, Amsterdam–New York 2001, pp. 233–267.
Elsakkers Marianne, *Genre hopping: Aristotelian criteria for abortion in Germania*, [in:] *Germanic Texts and Latin Models. Medieval Reconstructions*, ed. K. E. Olsen, A. Harbus, T. Hofstra, Luewen–Paris–Sterling, 2001, pp. 73–92.
Fleming Robin, *Bones for historians: Putting the body back into history*, [in:] *Writing Medieval Biography 750–1250. Essays in Honour of Professor Frank Barlow*, ed. D. Bates, J. Crick, S. Hamilton, Woodbridge 2006, pp. 27–48.
Ganshof François L., *Was waren die Capitularien?*, transl. W. Eckhardt, Weimar 1961.
Gasparri Stefano, *La memoria storica dei Longobardi*, [in:] *Le leggi die Longobardi*, ed. C. Azzara, S. Gasparri, Milano 1992, pp. V–XXII.
Goody Jack, *The Logic of Writing and the Organization of Society*, Cambrigde 1986.
Goody Jack, Watt Ian, *The consequences of literacy*, [in:] *Literacy in Traditional Society*, ed. J. Goody, Cambridge 1968, pp. 27–84.
Green Dennis H., *Language and History in Early Germanic World*, Cambridge 1998.
Günther Ludwig, *Ueber die Hauptstadien der geschichtlichen Entwicklung des Verbrechens der Körperverletzung und seiner Bestrafung*, Erlangen 1884.

Gysseling Maurits, *De germaanse woorden in de Lex Salica*, [in:] *Verslagen en Medelingen van de Koninklijke Akademie voor Nederlandse Taal-en Letterkunde, Nieuwe reeks*, Gent 1967, pp. 60–109.

Handwörterbuch zur deutschen Rechtsgeschichte, ed. A. Ekler, E. Kauffman, R. Schmidt-Wiegand, vol. I–III, Berlin 1971–1984.

Helten Willem van, *Zu den malbergischen glossen und salfränkischen formeln und ehnwörtern in der Lex Salica*, [in:] *Beiträge zur Geschichte der deutschen Sprache und Literatur*, ed. E. Sievers, vol. 25, Halle am Saale 1900.

His Rudolf, *Die Körperverletzungen im Strafrecht des deutschen Mittelalters*, ZRG GA 41, 1920, pp. 75–126.

Kern Hendrik, *Notes on the Frankish words in the Lex Salica*, [in:] *Lex Salica: the Ten Texts with the Glosses and the Lex Emendata*, ed. J. H. Hessels, London 1880, col. 429–564.

King Paul D., *King Chindasvind and the First Territorial Law-code of the Visigothic Kingdom*, [in:] *Visigothic Spain. New Approaches*, ed. E. James, Oxford 1980, pp. 131–157.

Krogmann Willy, *Entstehungszeit und Eigenart der Lex Frisionum*, Philologia Frisica, vol. 214 (1962), pp. 76–103.

Leeuw Gerardus van der, *Religion in Essence and Manifestation: A Study in Phenomenology*, transl. by J. E. Turner, New York 1963.

Le Goff Jacques, *Head or Heart? The political use of body metaphors in the Middle Ages*, [in:] *Fragment for the History of the Body*, ed. M. Feher, R. Naddaff, N. Tazi, vol. 3, New York 1989, pp. 13–26.

Le Goff Jacques, Truong Nicolas, *Une histoire du corps au Moyen Âge*, Paris 2003.

McKitterick Rosamond, *Carolingians and the Written Word*, Cambridge 1989.

McKitterick Rosamond, *Some Carolingian Law-Books and their function*, [in:] *Authority and Power. Studies on Medieval Law and Government Presented to Walter Ullman on his Seventieth Birthday*, ed. B. Tierney, P. Lienehan, Cambridge 1980, pp. 13–27.

McLeod Neil, *Parallel and paradox. Compensation in the legal systems of Celtic Ireland and Anglo-Saxon England*, Studia Celtica, vol. 16/17 (1981–1982), pp. 25–72.

McLeod Neil, *Compensation for fingers and teeth in early Irish law*, Peritia, vol. 16 (2002), pp. 344–359.

Modzelewski Karol, *Barbarzyńska Europa*, Warszawa 2004.

Modzelewski Karol, *Legem ipsam vetare non possumus. Królewski kodyfikator wobec potęgi zwyczaju*, [in:] *Historia, idee, polityka. Księga ofiarowana J. Baszkiewiczowi*, ed. F. Ryszka, Warszawa 1995, pp. 26–32.

Nehlsen Herman, *Zur Aktualität und Effektivität germanischer Rechtsaufzeichnungen*, [in:] *Recht und Schrift im Mittelalter*, ed. P. Classen, Sigmaringen 1977, pp. 449–502.

Nehlsen Herman, *Skalvenrecht zwischen Antike und Mittelalter. Germanisches und römisches Recht in der germanischen Rechtsaufzeichnugen I. Ostgoten, Westgoten, Franken, Langobarden*, Göttingen 1972.

Nehlsen Herman, *Der Grabfrevel in den germanischen Rechtsaufzeichnungen. Zugleich ein Beitrag zur Diskussion um Todesstrafe und Friedlosigkeit bei den Germanen*, [in:] *Zum Grabfrevel in vor- und frühgeschichtlicher Zeit. Untersuchungen zu Grabraub und „hangbrot" in Mittel- und Nordeuropa*, ed. H. Jankun, H. Nehlsen, H. Roth, Göttingen 1978, pp. 107–168.

Nehlsen Herman, *Lex Burgundionum*, [in:] *HRG*, vol. II, col. 1901–1915.

Nehlsen Herman, *Lex Visigothorum*, [in:] *HRG*, vol. II, col. 1966–1979.

Nehlsen Herman, *Lex Romana Burgundionum*, [in:] *HRG*, vol. II, col. 1927–1934.

Niederhellmann Anette, *Arzt und Heilkunde in der frühmittelalterlichen Leges*, Berlin 1983.

Niederhellmann Anette, *Heilkundiches in den Leges. Die Schadelverletzungen und ihre Bezeichnungen*, [in:] *Wörter und Sachen im Lichte der Bezeichnungsforschung*, ed. R. Schmidt-Wiegand, Berlin–New York 1981, pp. 74–90.

Nijdam Han, *Measuring wounds in the 'Lex Frisionum' and the old frisian register of fines*, Philologia Frisica, 1999, pp. 180–203.

O'Brien O'Keeffe Katherine, *Body and law in Anglo-Saxon England*, Anglo-Saxon England, 27 (1998), pp. 209–232.

Oliver Lisi, *The Beginnings of English Law*, Toronto 2002.

Oliver Lisi, *Body Legal in Barbarian Law*, Toronto 2011.

Ong Walter, *Orality and Literacy. The Thechnologizing of the Word*, London–New York 1982.

Ors Alvaro d', *El Codigo de Eurico. Estudios Visigoticos II*, Cuadernos del Instituto Juridico Espagñol, Rome–Madrid 1960.

Potkowski Edward, *Eschatologia germańska*, Euhemer – Przegląd Religioznawczy, 1963, no. 1 (32), pp. 25–39.

Potkowski Edward, *Dziedzictwo wierzeń pogańskich w średniowiecznych Niemczech. Defuncti vivi*, Warszawa 1973.

Rouche Michel, *A History of Private Life*, vol. 1: *From Pagan Rome to Byzantium*, transl. by A. Goldhammer, Cambrigde, Mass. 1992, pp. 453-518.

Rubin Stanley, *The Bot, or Composition in Anglo-Saxon law: Reassessment*, Journal of Legal History, vol. 17 (2) 1996, pp. 144–154.

Schmidt-Wiegand Ruth, *Zur Geschichte den malbergischen Glossen*, ZRG GA, vol. 74 (1957), pp. 220–231.

Schmidt-Wiegand Ruth, *Das frankische Wortgut der Lex Salica als Denkmal des Westfrnakischen*, Rheinische Vierteljahrblätter, vol. 33 (1969), pp. 396–422.

Schmidt-Wiegand Ruth, *Fränkische und frankolateinische Bezeichnungen für soziale Schichten und Gruppen in der Lex Salica*, Nachrichten der Akad. der Wissenschaften im Göttingen I, Phil.-hist. Klasse nr 4, Göttingen 1972, pp. 219–253.

Schmidt-Wiegand Ruth, *Lex Ribuaria*, [in:] *HRG*, vol. II, col. 1923–1927.

Schmidt-Wiegand Ruth, *Lex Salica*, [in:] *HRG*, vol. II, col. 1949–1962.

Schmidt-Wiegand Ruth, *Lex Thuringorum*, [in:] *HRG*, vol. II, col. 1965–1966.

Schmidt-Wiegand Ruth, *Stammesrecht und Volkssprache*, ed. D. Hüpper, Weinheim 1991.

Schmidt-Wiegand Ruth, *Spuren paganer Religiosität in frühmittelalterliche Rechtsquellen*, [in:] *Germanische Religionsgeschichte. Quellen und Quellenprobleme*, ed. H. Beck, D. Ellmers, K. Schier, Berlin–New York 1992, pp. 575–587.

Schmidt-Wiegand Ruth, *Spuren paganer Religiosität in den frühmittelalterlichen Leges*, [in:] *Iconologia Sacra. Mythos, Bildkunst und Dichtung in der Religions-Sozialgeschichte Alteuropas*, ed. H. Keller, N. Staubach, Berlin–New York 1994, pp. 249–262.

Schmidt-Wiegand Ruth, *Wargus. Eine Bezeichnung für den Unrechtstäter in ihrem wort-geschichtlichen Zusammenhang*, [in:] *Zum Grabfrevel in vor- und frühgeschichtlicher Zeit. Untersuchungen zu Grabraub und „haugbrot" in Mittelund Nordeuropa*, ed. H. Jankuhn, H. Nehlsen, H. Roth, Göttingen 1978, pp. 188–196.

Schmidt-Wiegand Ruth, *Christentum und pagane Religiosität in Pactus und Lex Alamannorum*, [in:] *Die Alemannen und das Christentum*, ed. S. Lorenz, B. Scholkmann, Leitfelden–Echtringen 2003, pp. 113–124.

Schott Clausdieter, *Pactus, Lex und Recht*, [in:] *Die Alemannen in der Frühzeit*, ed.W. Hübner, Bühl–Baden 1974, pp. 135–168.

Schott Clausdieter, *Lex Alamannorum*, [in:] *HRG*, vol. II, col. 1879–1886.

Schott Clausdieter, *Der Stand der Leges-Forschung*, Frühmittelalterliche Studien, vol. 13, 1979, pp. 29–55.

Schott Clausdieter, *Zur Geltung der Lex Alamannorum*, [in:] *Die historische Landschaft zwischen Lech und Vogesen; Forschungen und Fragen zur gesamtalemannischen Geschichte*, ed. P. Fried, W.-D. Sick, Freiburg 1988, pp. 75–105.

Schott Clausdieter, *Die Leges Alamannorum*, [in:] *Lex Alamannorum. Das Gesetz der Alemannen, Text – Übersetzung – Kommentar zum Faksimile aus der Wandalgarius-Handschrift Codex Sangallensis 731*, Augsburg 1993, pp. 1–24.
Schwanenflügel Sigfrid von, *Die Körperverletzung in der ersten geschribenen Rechten der Germanen (etwa 500–1300 n. Chr.)*, Göttingen 1950.
Sellert Wolfgang, *Aufzeichnung des Recht und Gesetz*, [in:] *Das Gesetz im Spatantike und frühem Mittelalter*, ed. W. Sellert (Abhandlungen der Akademie des Wissenschaften in Gottingen 196), Göttingen 1992, pp. 67–102.
Siems Harald, *Studien zur Lex Frisionum*, Ebelsbach am Main 1980.
Siems Harald, *Lex Baiwariorum*, [in:] *HRG*, vol. II, col. 1887–1901.
Siems Harald, *Lex Frisionum*, [in:] *HRG*, vol. II, col. 1916–1922.
Siems Harald, *Lex Romana Visigothorum*, [in:] *HRG*, vol. II, col. 1939–1949;
Sigurðsson Gísli, *The Medieval Icelandic Saga and Oral Tradition. A Discourse on Method*, transl. N. Jones, Cambridge, Mass.– London 2004.
Simpson Walter S., *The Laws of Ethelberht*, [in:] *On the Laws and Customs of England. Essays in Honor of Samuel E. Thorne*, ed. M. S. Arnold, Th. A. Green, S. A. Scully, St. D. White, Chapell Hill 1981, pp. 3–17.
Smith Julia M. H., *Europe after Rome. A New Cultural History 500-1000*, Oxford 2005.
Snell Bruno, *The Discovery of the Mind in Greek Philosophy and Literature*, transl. T. G. Rosenmeyer, New York 1982.
The Burgundian Code. Book of Constitutions or Law of Gundobad. Additional Enactments, transl. K. Fischer Drew, Philadelphia 1972.
Theurkauf Gerhard, *Lex, Speculum, Compendium iuris. Rechtsaufzeichnung und Rechtsbewußtsein Norddeutschland vom 8. bis 16. Jahrhundert*, Köln– Graz 1968.
Thöne Franz, *Die namen der menschlischen körperteile bei den Angelsachsen*, Kiel 1912.
Tyrrell Andrew, *'Corpus Saxonum': Early Medieval bodies and corporeal identity*, [in:] *Social Identity in Early Medieval Britain*, ed. W. O. Frazer, A. Tyrrell, London–New York 2000, pp. 137–155.
Vollrath Hanna, *Gesetzgebung und Schriftlichkeit. Das Beispiel der angelsächsischen Gesetze*, Historische Jahrbuch, vol. 99 (1979), pp. 28–54.
Wallace-Hadrill John, *The Long-Haired Kings*, London 1962.
Wallace-Hadrill John, *Early Germanic Kingship in England and on the Continent*, Oxford 1971.
Wilda Wilhelm E., *Das Strafrecht der Germanen*, Aalen 1960 [1842].

Wood Ian, *Administration, law and culture in Merovingian Gaul*, [in:] *The Uses of Literacy in Early Medieval Europe*, ed. R. McKitterick, Cambridge 1990, pp. 63–81.
Wood Ian, *The Code in Merovingian Gaul*, [in:] *The Theodosian Code*, ed. J. Harries, I. Wood, New York 1993, pp. 161–177.
Wood Ian, *The Merovingian Kingdoms, 450–751*, London–New York 1994.
Wormald Patrick, '*Lex Scripta' and 'Verbum Regis': Legislation and Germanic Kingship, from Euric to Cnut*, [in:] *Early Medieval Kingship*, sed. P. H. Sawyer, I. N. Wood, Leeds 1977, pp. 105–138.
Wormald Patrick, „*Inter cetera bona... genti suae': law-making and peacekeeping in the earliest English kingdoms*, [in:] *La giustizia nell'alto medioevo (secoli V–VIII)*, vol. II, ed. C. Leonardi, Spoleto 1995, pp. 963–996.
Wormald Patrick, *Exempla Romanorum: the Earliest English legislation in context*, [in:] *Rome and the North*, ed. A. Ellegård, G. Åkerström- Hougen, Göteborg 1996, pp. 15–27.
Wormald Patrick, *The making of English law: King Alfred to the Twelfth Century*, vol. 1, *Legislation and its Limits*, Oxford 1999.
Wormald Patrick, *The* Leges Barbarorum*: law and ethnicity in the post-roman West*, [in:] *Regna and Gentes. The Relatioship between Late Anique and Early Medieval Peoples and Kingdoms in the Transformation of the Roman World*, eds. H.-W. Goetz, J. Jarnut, and W. Pohl, Leiden-Boston 2003.
Zeumer Karl, *Geschichte der westgotischen Gesetzgebung I*, Neues Archiv, vol. 23 (1898) pp. 419–516, 24 (1899) pp. 39–122 and 571–630, 26 (1901) pp. 91–149.

www.ingramcontent.com/pod-product-compliance
Ingram Content Group UK Ltd.
Pitfield, Milton Keynes, MK11 3LW, UK
UKHW041912140426
5217IPUK00002B/14